XXXIV Symposium on Nuclear Physics 2011

Journal of Physics: Conference Series Volume 322

Morelos, Mexico
4-7 January 2011

Editors:

Libertad Barron-Palos **Roelof Bijker**

ISBN: 978-1-61839-292-3
ISSN: 1742-6588

Printed from e-media with permission by:

Curran Associates, Inc.
57 Morehouse Lane
Red Hook, NY 12571

Printed by Curran Associates, Inc. (2011)

For permission requests, please contact the Institute of Physics
at the address below.

Institute of Physics
Dirac House, Temple Back
Bristol BS1 6BE UK

Phone: 44 1 17 929 7481
Fax: 44 1 17 920 0979

techtracking@iop.org

Additional copies of this publication are available from:

Curran Associates, Inc.
57 Morehouse Lane
Red Hook, NY 12571 USA
Phone: 845-758-0400
Fax: 845-758-2634
Email: curran@proceedings.com
Web: www.proceedings.com

TABLE OF CONTENTS

PREFACE

In the present volume of the *Journal of Physics: Conference Series* we publish the proceedings of the 'XXXIV Symposium on Nuclear Physics', which was held from 4–7 January 2011 at the Hacienda Cocoyoc, Morelos, Mexico. The proceedings consist of 19 contributions that were presented as invited talks at the meeting. The abstracts of all contributions, plenary talks and posters were published in the Conference Handbook.

The Symposium on Nuclear Physics has a long and distinguished history. From the beginning it was intended to be a relatively small meeting designed to bring together some of the leading nuclear scientists in the field. Its most distinctive feature is to provide a forum for specialists in different areas of nuclear physics, both theorists and experimentalists, students, postdocs and senior scientists, in a relaxed and informal environment providing them with a unique opportunity to exchange ideas. From the first meeting in Oaxtepec in 1978, the Symposium has been organized every year without interruption, which makes the present Symposium the 34th in a row.

The scientific program consisted of 27 invited talks and 17 posters on a wide variety of hot topics in contemporary nuclear physics, ranging from the traditional fields of nuclear structure (Draayer, Pittel, Van Isacker, Fraser, Lerma, Cejnar, Hirsch, Stránský and Rath) and nuclear reactions (Aguilera, Gómez-Camacho, Scheid, Navrátil and Yennello) to radioactive beams (Padilla-Rodal and Galindo-Uribarri), nuclear astrophysics (Aprahamian, Civitarese and Escher), hadronic physics (Bijker, Valcarce and Hess), fundamental symmetries (Liu, Barrón-Palos and Baessler) and LHC physics (Menchaca-Rocha and Paic). The high quality of the talks, the prestige of the speakers and the broad spectrum of subjects covered in the meeting, shows that nuclear physics is a very active area at the frontier of scientific research which establishes bridges between many different disciplines.

Libertad Barrón-Palos
Roelof Bijker
Editors

XXXIV SYMPOSIUM ON NUCLEAR PHYSISCS
Cocoyoc Morelos. January 4-7 2011

Journal of Physics: Conference Series **322** (2011) 012001 doi:10.1088/1742-6596/322/1/012001

Ab Initio Symmetry-Adapted No-Core Shell Model

J P Draayer, T Dytrych, and K D Launey

Department of Physics and Astronomy, Louisiana State University, Baton Rouge, LA 70803, USA

E-mail: draayer@lsu.edu

Abstract. A multi-shell extension of the Elliott SU(3) model, the SU(3) symmetry-adapted version of the no-core shell model (SA-NCSM), is described. The significance of this SA-NCSM emerges from the physical relevance of its SU(3)-coupled basis, which – while it naturally manages center-of-mass spuriosity – provides a microscopic description of nuclei in terms of mixed shape configurations. Since typically configurations of maximum spatial deformation dominate, only a small part of the model space suffices to reproduce the low-energy nuclear dynamics and hence, offers an effective symmetry-guided framework for winnowing of model space. This is based on our recent findings of low-spin and high-deformation dominance in realistic NCSM results and, in turn, holds promise to significantly enhance the reach of *ab initio* shell models.

1. Introduction

A quintessential goal of theoretical nuclear physics is to provide a detailed explanation of the properties of the atomic nucleus based on knowledge of the strong force between its constituents. When this goal is achieved, it will be possible to predict structures of exotic nuclei and reactions that take place in extreme environments – from the interiors of stars to the core of nuclear reactors. With a view toward this goal, the *ab initio* symmetry-adapted no-core shell model (SA-NCSM) aims to expand dramatically the reach as well as the impact of current *ab initio* shell models toward accommodating heavier mass nuclei together with highly deformed and cluster sub-structures. By effectively ferreting out important interactions at the fundamental quantum level, the SA-NCSM holds promise to achieve predictions of nuclear properties with unprecedented accuracy and provide a deeper understanding of the underlying forces.

Currently, the no-core shell model (NCSM) is used to reproduce features of the deuteron up through ^{16}O and even for some selected cases beyond (e.g., see Refs. [1, 2] for NCSM and cluster NCSM/RGM results). The symmetry-adapted no-core shell model (SA-NCSM) approach augments the NCSM concept by recognizing that the choice of coordinates, especially when deformed nuclear shapes dominate, is crucial and presents a solution in terms of coordinates that reflect symmetries inherent to nuclear systems. While the SA-NCSM states can be obtained through a unitary transformation from the m-scheme basis used in the NCSM, the growth of the model space within the SA-NCSM framework can be managed by winnowing to only physically relevant states as determined through symmetry considerations.

Journal of Physics: Conference Series **322** (2011) 012001 doi:10.1088/1742-6596/322/1/012001

2. Significance of an *ab initio* symmetry-adapted no-core shell model

The underlying symmetry of the SA-NCSM is the SU(3) group, a subgroup of the physically relevant symplectic Sp(3, $\mathbb{R}$) group, which underpins a shell model that is known to provide a microscopic formulation of the famous Bohr-Mottelson collective model and to be a multiple oscillator shell generalization of the successful Elliott SU(3) model [3]. Indeed, the significance of the symplectic symmetry for a microscopic description of a quantum many-body system of interacting particles [3] naturally emerges from the physical relevance of its 21 generators, which are directly related to the particle momentum and coordinate operators and realize important observables, such as the many-particle kinetic energy, the mass quadrupole moment and angular momentum operators, together with multi-shell collective vibrations and vorticity degrees of freedom for a description from irrotational to rigid rotor flows. The translationally invariant (intrinsic) SU(3) generators can be written in terms of the harmonic oscillator raising ($b^{\dagger(1\,0)}$) and lowering ($b^{(0\,1)}$) operators,

$$C^{(1\,1)}_{\mathfrak{L}M} = \sqrt{2}\sum_i \left[b_i^\dagger \times b_i\right]^{(1\,1)}_{\mathfrak{L}M} - \frac{\sqrt{2}}{A}\sum_{s,t}\left[b_s^\dagger \times b_t\right]^{(1\,1)}_{\mathfrak{L}M}, \quad (1)$$

where the sums are over all A particles of the system. The eight operators $C^{(1\,1)}_{\mathfrak{L}M}$ generate the SU(3) subgroup of Sp(3, $\mathbb{R}$) and are related to the angular momentum operator L_{1q} $(= C^{(1\,1)}_{1q}$, $q = 0, \pm 1)$ and the Elliott algebraic quadrupole moment tensor $\mathcal{Q}^a_{2q}$ $(= \sqrt{3}C^{(1\,1)}_{2q}$, $q = 0, \pm1, \pm2)$. The deformation-related $(\lambda\,\mu)$ set of quantum numbers labels SU(3) irreducible representations, irreps. Consequently, SU(3)-symmetric states (and hence symplectic basis states that are built on these) bring forward important information about nuclear shapes and deformation in terms of $(\lambda\,\mu)$, for example, $(0\,0)$, $(\lambda\,0)$ and $(0\,\mu)$ describe spherical, prolate and oblate shapes, respectively. Using conventional labels, the SU(3)×SU$_S$(2) basis of the SA-NCSM is given as, $|\alpha(\lambda\,\mu)\kappa(LS)JM_J\rangle$ with α labeling the many-body configuration and κ distinguishing multiple occurrences of the same L value in the parent configuration (S and J denote spin and total angular momentum, respectively). This basis can be obtained as a unitary transformation from the NCSM m-scheme basis and hence, spans the entire space.

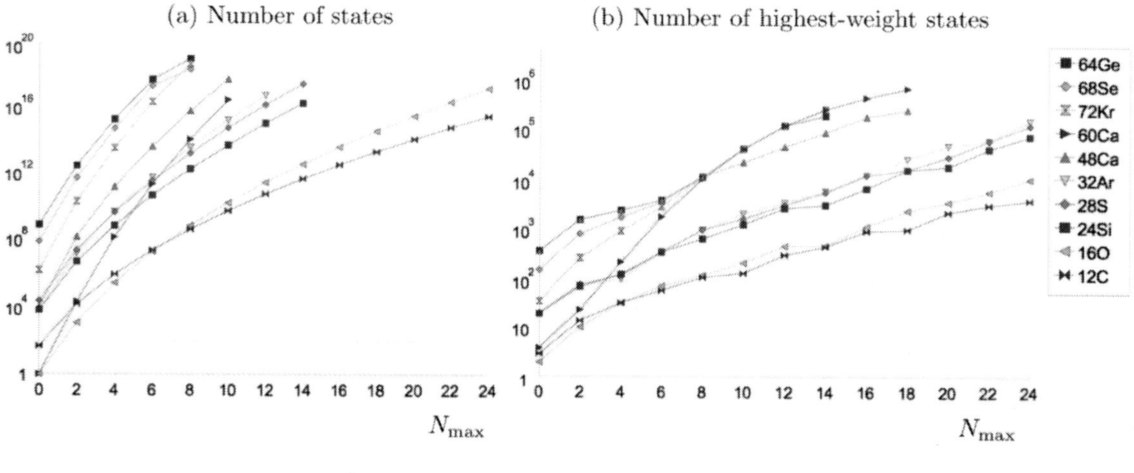

Figure 1. Number of (a) NCSM many-body states and (b) all single-shell SU(3) × SU(2) highest-weight states of the SA-NCSM many-body model space as a function of the $N_{\max}$ many-body cutoff for several representative nuclei in the p, sd, and pf shells.

The SA-NCSM implements a fast method for calculating matrix elements of arbitrary two-body operators in the SU(3)×SU$_S$(2) symmetry-adapted basis. This facilitates both the

Journal of Physics: Conference Series **322** (2011) 012001 doi:10.1088/1742-6596/322/1/012001

evaluation of the Hamiltonian matrix elements and the use of the resulting eigenvectors to evaluate other experimental observables. Any two-body operator can be expanded in terms of SU(3)-coupled fermion creation and annihilation operators,

$$\left[\{a^\dagger_{\eta_i} \otimes a^\dagger_{\eta_j}\}^{(\lambda_{of}\,\mu_{of})}_{S_{of}} \otimes \{a_{\eta_l} \otimes a_{\eta_k}\}^{(\mu_{oi}\,\lambda_{oi})}_{S_{oi}}\right]^{\alpha_o(\lambda_o\,\mu_o)}_{\kappa_o L_o S_o J_o M_{oJ}}. \tag{2}$$

Here $a^\dagger_\eta$ and a_η denote fermion creation and annihilation operators in the η major oscillator shell with SU(3) symmetry $(\eta\, 0)$ and $(0\,\eta)$, respectively. It is a straightforward matter to generalize this expansion for a three-body operator. Exploiting the SU(3)$\times$SU$_S$(2) symmetry of the SA-NCSM basis states, it suffices to evaluate the following SU(3)-reduced matrix elements, r.m.e. ('i' for 'initial', 'f' for 'final', 'o' for 'operator'):

$$\langle \alpha_f(\lambda_f\,\mu_f)S_f ||| \left[\{a^\dagger_{\eta_i} \otimes a^\dagger_{\eta_j}\}^{(\lambda_{of}\,\mu_{of})}_{S_{of}} \otimes \{a_{\eta_l} \otimes a_{\eta_k}\}^{(\mu_{oi}\,\lambda_{oi})}_{S_{oi}}\right]^{\alpha_o(\lambda_o\,\mu_o)}_{S_o} ||| \alpha_i(\lambda_i\,\mu_i)S_i\rangle. \tag{3}$$

As the SU(3) configurations in the SA-NCSM are constructed by coupling a sequence of the physically allowed single-shell SU(3) representations, the reduced matrix elements (3) can be extracted from the full matrix elements between the highest-weight states of the single-shell SU(3) representations (a SU(3) highest-weight state, which is annihilated by the raising SU(3) generators, is analogous to a $|J, M_J = J\rangle$ state for the case of the SU(2) group).

This factorization of matrix elements into the product of single-shell SU(3) r.m.e. and the associated SU(3) coupling coefficient, which can be calculated 'on the fly', reduces the number of key pieces of information required to the single-shell SU(3) r.m.e., and these track with the number of single-shell SU(3) highest-weight states. Clearly, compared to the explosive growth of the number of NCSM many-body configurations with increasing $N_{\max}$ cutoff (Fig. 1a), the number of single-shell highest-weight states of the SA-NCSM (Fig. 1b) is manageable even, e.g., for 'pf-shell' nuclei.

Furthermore, within the SA-NCSM framework, the growth of the model space can be managed by winnowing the model space to only physically relevant states as determined through symmetry considerations. The underlying concept of this framework is illustrated in our proof-of-principle study [4, 5] that exploits symplectic Sp(3, $\mathbb{R}$) symmetry and its SU(3) subgroup symmetry in an analysis of large-scale nuclear physics applications for ^{12}C and ^{16}O. What one learns from the outcome of these studies is that typically a small fraction of the full model space, several orders of magnitude less than that of the corresponding NSCM approach, suffices to represent most of the physics – typically 90% or more as measured by projecting NCSM results onto a symmetry-adapted equivalent basis and noting that only a small subset of the full space contributes to the low-energy dynamics.

3. Low-spin and high-deformation dominance

The symmetry-guided framework of the SA-NCSM utilizes a natural preference towards low-spin and high-deformation dominance as revealed in realistic NCSM wavefunctions. Specifically, our recent study [6] showed that for the $6\hbar\Omega$ NCSM results for ^{12}C using the effective N^{3}LO interaction (and similarly for other interaction choices, such as JISP16) with $\hbar\Omega$=15 MeV (Fig. 2), proton (neutron) spin values S_π $(S_\nu) = 0$ and 1 are sufficient to describe about 99% of the converged $J = 0^+_{\rm gs}$, 2^+_1, and 4^+_1 ^{12}C NCSM eigenvectors. The residual 1% involves time/memory consuming higher-spin and consequently less relevant configurations. Similarly, contributions from high S_π (S_ν) proton (neutron) spin values are found negligible for the $J = 1$ state of ^{12}C shown in Fig. 3 for two JISP16 and N^{3}LO effective interactions for $\hbar\Omega$=15 MeV. Hence, retaining only the relevant proton (neutron) spin values allows one, with the same

Journal of Physics: Conference Series **322** (2011) 012001 doi:10.1088/1742-6596/322/1/012001

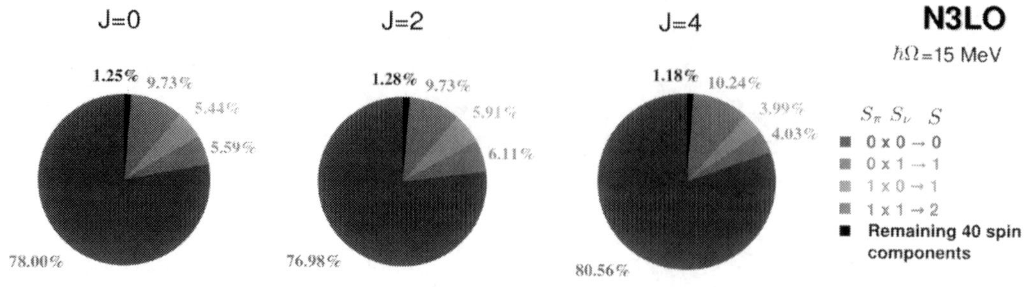

Figure 2. Intrinsic spin structure of $J = 0^+_{\rm gs}, 2^+_1$, and 4^+_1 NCSM states in ^{12}C obtained using the N^{3}LO effective interaction in the $N_{\rm max} = 6$ model space with $\hbar\Omega = 15$ MeV.

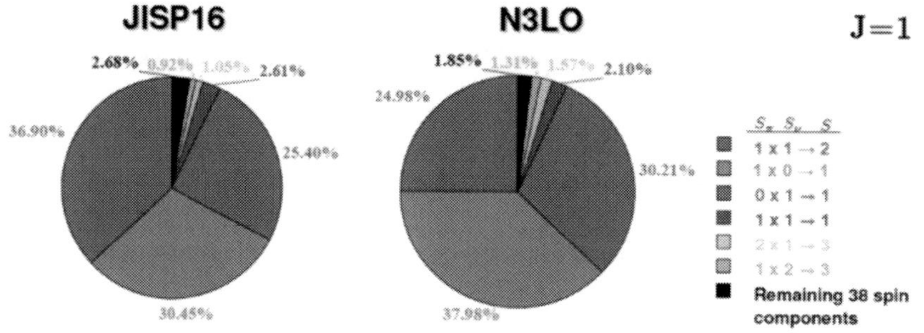

Figure 3. Intrinsic spin structure of $J = 1$ NCSM states in ^{12}C obtained using JISP16 and N^{3}LO effective interactions in the $N_{\rm max} = 6$ model space with $\hbar\Omega = 15$ MeV.

computer resources, to accommodate the full basis within the selected spin spaces up through higher $\hbar\Omega$ values.

Furthermore, in a proton-neutron formalism, the spin quantum numbers dictate, within a single shell, the spatial symmetry because the fermion statistics requires overall antisymmetry within each sector. Hence, low-spin implies maximum spatial symmetry, which is where one finds the SU(3) configurations of maximum spatial deformation. Indeed, for the 0^+ ground state in ^{12}C, calculated by NCSM with JISP16 interaction for $\hbar\Omega = 15$ MeV [5], the SU(3) states that correspond to the most deformed oblate shape realize the major component of this state (Fig. 4).

4. Symmetry-guided framework

Winnowing considerations of the type shown above illustrate that the SA-NCSM offers a systematic framework for down-selecting to physically relevant and manageable subspaces associated with full NCSM based on spin and deformation selection, which are complementary and mutually reinforcing. This is illustrated in Fig. 5, where a manageable (e.g, $N_{\rm max} = 6$) NCSM many-body space is schematically shown (a) together with an associated manageable space within the symmetry-guided framework that underpins the SA-NCSM (b). In the SA-NCSM structured approach, the space is first separated into proton and neutron subspaces and then, in keeping with the LS coupling foundation of the SA-NCSM, this is further organized according to the total proton spin (S_ν) and neutron spin (S_π) coupled to total spin (S). This

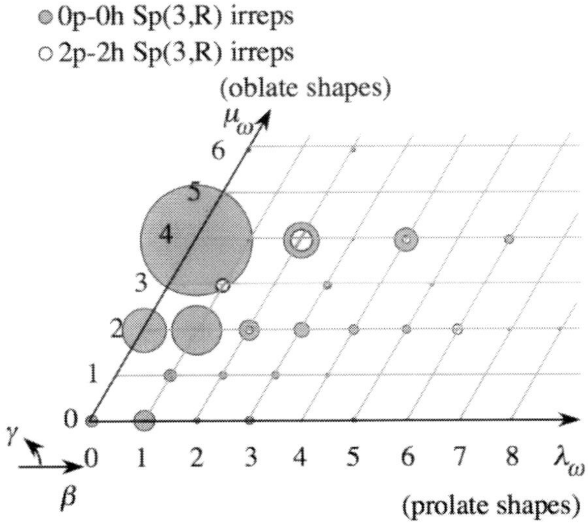

Figure 4. Probabilities (specified by the area of the circles) for the symplectic states which make up the most important 0-particle-0-hole (0p-0h) (blue circles) and $2\hbar\Omega$ 2p-2h (empty circles) symplectic irreps, within the 0^+ ground state in ^{12}C, calculated by NCSM with JISP16 interaction, $\hbar\Omega = 15$ MeV. The Sp(3, $\mathbb{R}$) states are grouped according to their (λ_ω μ_ω) SU(3) symmetry, which is mapped onto the (β γ) shape variables of the collective model.

decomposition enables one to trim the space to the dominant low-spin configurations.

These spin spaces can be further organized into SU(3) structures, each of which realizes a nuclear shape deformation (represented as 'ellipsoids' in Fig. 5). Since typically configurations of maximum spatial deformation dominate, only a percentage of the proton(neutron)/spin subspace need be further considered (green 'ellipsoids') to accommodate the physically most significant configurations, including particle-hole excitations that are important for a description of clustering modes, in higher $\hbar\Omega$ model spaces (e.g., $N_{\max} = 10 - 12$ in Fig. 5). And beyond these, the Sp(3, $\mathbb{R}$) symplectic symmetry can guide an extension of model spaces to incorporate even higher $\hbar\Omega$ values (e.g., $N_{\max} = 14 - 16$ in Fig. 5) [4] required to gain convergence of the lowest bound 0^+ states in light nuclei (green cones in Fig. 5).

An illustrative example for the symmetry-guided framework of the SA-NCSM is shown in Fig. 6 for the case of the $N \sim Z$ pf-nucleus ^{64}Ge of astrophysical significance. The figure shows the combinatorial (near-exponential) growth of the full NCSM space (blue squares). Employing spin considerations, the full space can be reduced roughly by an order of magnitude while achieving considerable decrease in computing intensity because low-spin configurations typically have a simpler structure than for higher ones. In the example of Fig. 6, we chose to retain only proton (neutron) spin values $S_\pi = 0$ and 1 ($S_\nu = 0$ and 1) coupled to total spin $S = 0$, 1 and 2 (no restrictions, red diamonds). Further reductions of several order of magnitude can be achieved by selecting SU(3) proton (neutron) modes according to their shape deformation with the most deformed structures playing the foremost role in nuclear collectivity. As for the spin, the selected proton SU(3) configurations are coupled to the selected neutron configurations to yield the total number of many-body states (yellow, green and purple triangles). Beyond these winnowing considerations, only reduced matrix elements of the associated highest-weight states need to be calculated and stored because the associated coupling coefficients can be computed 'on the fly'.

For shell models that employ restricted model spaces a suitable Hamiltonian renormalization,

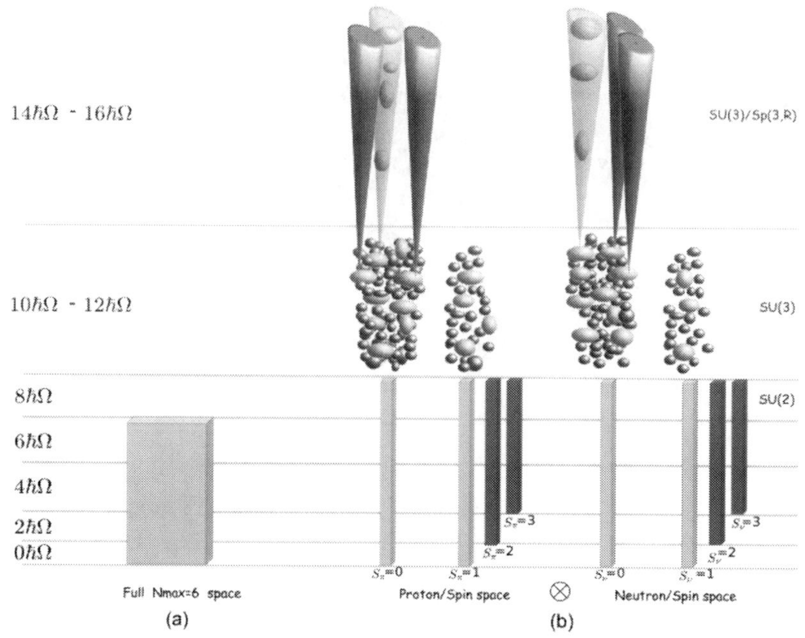

Figure 5. Schematic illustration of a manageable many-body basis of (a) NCSM ($N_{\max} = 6$) and (b) SA-NCSM with the symmetry-guided extension of high-$\hbar\Omega$ configurations. Green (light) color denotes physically significant configurations and grey (dark) color denotes physically irrelevant configurations.

which is properly derived from realistic nucleon-nucleon interactions tied to QCD, is required in order to preserve predictive capabilities. A novel approach, the Similarity Renormalization Group (SRG) [7], which was recently adopted in nuclear physics and continues to enjoy successes in, e.g, condensed matter and high energy physics, decouples spaces that are relevant for the regime of nuclear dynamics from highly-energetic irrelevant configurations. Moreover, reflecting the symmetries in play for the SA-NCSM, we suggest a SU(3)-based SRG renormalization [8], which is suitable for SU(3)-adapted shell model calculations. This is because the SRG in a SU(3) basis results in a unitarily transformed NN interaction that respects the SU(3) symmetry, that is, a (near-)diagonal Hamiltonian in the basis used. In addition, while an important but challenging part of the SRG approaches is to properly account for the many-body forces induced during the renormalization (in order to preserve the unitarity of the SRG transformations), the use of symmetries, e.g. highest-weight states and reduced matrix elements, plays a crucial role in addressing this issue.

In short, the SA-NCSM builds upon a proton-neutron formalism in a LS coupling scheme with the proton/neutron spaces organized into subspaces of definite spin and further into spatial SU(3) representations that key to deformation, and for even higher $\hbar\Omega$ model spaces, into symplectic $\mathrm{Sp}(3,\mathbb{R})\supset\mathrm{SU}(3)$ structures. The symmetry-guided framework utilizes the physical relevance at each level of the structured space, namely, low-spin and high-deformation dominance together with most important patterns of symplectic excitations. This approach will open up an entire region of the periodic table to investigation with *ab initio* methods with forefront predictive capabilities.

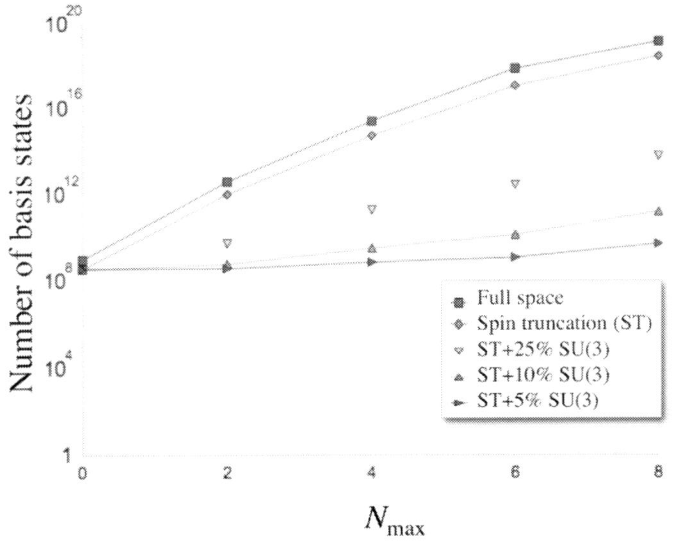

Figure 6. Dimensions of various model spaces for ${}^{64}_{32}Ge_{32}$: conventional NCSM model spaces (blue squares) and reduced model space dimensions when only proton (neutron) S_π (S_ν) = 0 and 1 values coupled to total spin 0, 1 and 2 are considered (red diamonds) and when beyond this only 25%, 10% and 5% of the most deformed proton (neutron) SU(3) configurations are retained (yellow, green and purple triangles, respectively) as a function of N_{max}.

5. Conclusion

The SA-NCSM advances an extensible microscopic framework for studying nuclear structure and reaction processes – strong as well as weak interaction dominated, that capitalizes on advances being made in *ab initio* methods while exploiting symmetries – exact and partial, known to dominate the dynamics. We have developed a symmetry-adapted shell model with the view toward exploring the properties of nuclei far from stability using externally provided realistic interactions derived from Quantum Chromodynamics (QCD) considerations. The method is applied first for light systems so the currently available *ab initio* methods can be used to guide the development, but ultimately pushing towards heavier ones that will require symmetry guided winnowing decisions.

Supported by the U.S. National Science Foundation (PHY-0500291 & OCI-0904874), the U.S. Department of Energy (DE-SC0005248), and the Southeastern Universities Research Association (http:/www.sura.org).

References

[1] Navrátil P, Vary J and Barrett B 2000 *Phys. Rev. Lett.* **62** 054311
[2] Navrátil P, Quaglioni S, Stetcu I and Barrett B 2009 *J. Phys. G: Nucl. Part.* **36** 083101
[3] Rosensteel G and Rowe D 1977 *Phys. Rev. Lett.* **38** 10
[4] Dytrych T, Sviratcheva K, Bahri C, Draayer J and Vary J 2007 *Phys. Rev. Lett.* **98** 162503
[5] Dytrych T, Sviratcheva K D, Draayer J P, Bahri C and Vary J P 2008 *J. Phys. G: Nucl. Part. Phys.* **35** 123101
[6] Dytrych T, Draayer J P, Sviratcheva K D, Bahri C and Vary J P 2009 *"Nuclear Structure And Dynamics '09": Proceedings of the International Conference, Dubrovnik, Croatia, 4-8 May 2009* vol 1165 ed Milin M, Niksic T, Szilner S and Vretenar D (AIP Conference Proceedings)
[7] Bogner S, Furnstahl R and Perry R 2007 *Phys. Rev. C* **75** 061001(R)
[8] Launey K, Dytrych T and Draayer J 2010 *submitted*

Journal of Physics: Conference Series **322** (2011) 012002 doi:10.1088/1742-6596/322/1/012002

Extended application of the DMRG methodology to nuclei in the $1f - 2p$ shell

S Pittel[1] **and B Thakur**[1,2]

[1]Bartol Research Institute and Department of Physics and Astronomy, University of Delaware, Newark, DE 19716, USA

[2]LONI Institute and Center for Computational Technology, Louisiana State University, Baton Rouge, LA 70803-4001 USA

E-mail: pittel@bartol.udel.edu

E-mail: bthakur@cct.lsu.edu

Abstract. In this paper, we present systematic results of the J-DMRG method (an angular-momentum-conserving version of the Density Matrix Renormalization Group) in truncated shell-model calculations of nuclei in the $1f-2p$ shell. In contrast to earlier studies, these calculations consider both even-even and odd-mass nuclei. Comparison of the results of these calculations with those of exact shell-model treatments of the same nuclei shows that the method converges smoothly to the exact results in all cases and furthermore that the fraction of the full shell-model space required to obtain accurate agreement with the exact results decreases rapidly as the size of the full space grows. This latter result bodes well for the future applicability of the method to even larger nuclei with even larger shell-model spaces as is the ultimate goal of our development of this approach to large-scale shell model studies.

1. Introduction

The last few decades have witnessed a rapid growth in the capabilities of the Configuration Mixing Shell Model to treat complex atomic nuclei. Large-scale calculations with exact diagonalization can now be carried out for most of the nuclei in the $1f - 2p$ shell using sophisticated shell-model codes such as CMichSM, MSHELL, NATHAN, RedStick and the publicly available ANTOINE. Ideally one would like to extend the scope of shell-model calculations to even heavier nuclei. But with an exponentially increasing configuration space, any computational improvements realized in the foreseeable future might still be insufficient for larger calculations beyond the $1f-2p$ shell. Over the years, a variety of approximate truncation strategies have been developed to address this limitation of the shell-model algorithm. One such method is provided by the Density Matrix Renormalization Group (DMRG), a method first introduced [1] for the study of quantum lattices and subsequently extended to finite Fermi systems [2]. In this context it has been suggested as a possible approach for use with the nuclear shell model. In this work, we focus on an issue crucial to its possible applicability to larger nuclear systems, how the method scales with the size of the shell-model problem.

2. Background and Motivation

The DMRG is a method for systematically including the degrees of freedom of a many-body problem. The algorithm begins by partitioning the complete space of the problem in terms of lattice sites, with each site hosting a small fraction of the complete space. In the nuclear context, these sites represent single-particle orbitals. The basic idea of the method is to systematically grow the system to include an additional site. At each stage of the iterative growth procedure, a truncation of the states in this enlarged system (called a block) is implemented through the use of density matrix considerations based on their importance within the complete space. The method is constructed so that the optimal set of states within each block is iteratively updated by sweeping over the various sites until convergence is achieved. For a comprehensive review of the the DMRG procedure, we refer the reader to ref. [2].

The DMRG when applied to atomic nuclei has somewhat unique features, especially when compared with its more typical applications both to quantum lattices and in quantum chemistry. On the one hand, the fact that the lattice sites appropriate for nuclear applications are single-particle orbitals implies that they can have varying sizes, and this leads to computational difficulties in their iterative treatment. A second key difference is the presence of two kinds of fermions, making the choice of an appropriate order in which to iteratively include the associated orbitals more challenging. Our earlier applications of the DMRG methodology have addressed these issues and have come up with an optimal algorithm for their treatment. The algorithm we have adopted, the J-DMRG, also has the feature that it conserves angular momentum throughout the iterative procedure. A description of the algorithm we have developed and the associated terminology can be found in refs. [3, 4].

In our earlier applications of the J-DMRG method [5, 6], we focussed on two specific even-even nuclei in the $1f-2p$ shell, ^{48}Cr and ^{56}Ni. Those calculations indicated that the J-DMRG method indeed converges smoothly to the exact results for both nuclei and furthermore that the method seems to be scaling well with the size of the problem. In the current work, our goal is to see whether those results carry over to a broader range of nuclei and furthermore to better assess the scalability of the method. Towards that end we will discuss the application of the DMRG method to a wide range of nuclei in the $1f-2p$ shell, both even-even and with odd mass. We will assess how rapidly the method converges for these different types of nuclei and also how scalable is the method as a function of the size of the full shell-model problem.

In the course of the study, we will also address other issues that were hinted at by our earlier investigations. One concerns which state or states of the full space to use in implementing the density matrix truncation algorithm. It is typical to focus on just the ground state and to build a reduced density matrix derived from that state only. Here we address the issue of whether it is better to include information on selected excited states as well by introducing mixed density matrices to define the optimal structure of each block.

3. General features of our calculational apparatus

All of the calculations to be described later are for nuclei in the $1f-2p$ shell. In all, we assume a description in terms of valence neutrons and protons outside a doubly-magic ^{40}Ca core and restricted to the orbitals of that one major shell. The effective hamiltonian we use is based on the GXPF1A effective two-body interaction first proposed in reference [7], with the same set of single-particle energies as used in that work as well. This effective hamiltonian is known to provide a somewhat better description of nuclei in this region than the KB3 hamiltonian [8] used in our earlier investigations.

In all the calculations reported here, we use the so-called three-block growth strategy adopted in our earlier studies. The key features of this strategy are that (*i*) when we order our single-particle orbitals into a chain, we keep those for neutrons on one side of the chain and those for protons on the other side, and (*ii*) as we grow the system we only build blocks from orbitals of the same type. It was found in our earlier studies that this growth strategy has significant computational advantages to the more-traditional two-block approach in which mixed neutron-proton blocks would be constructed as well.

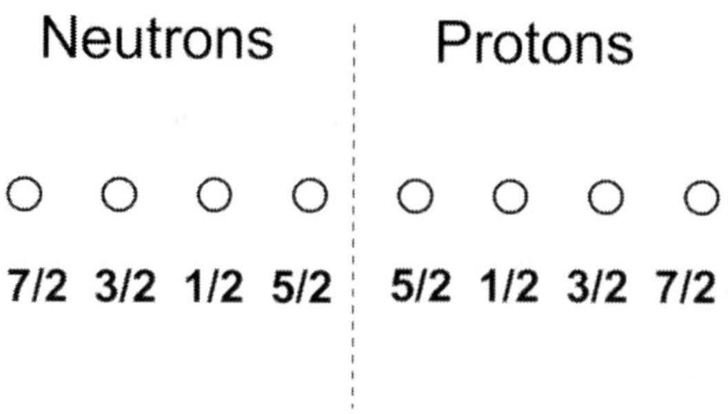

Figure 1. Order of single-particle levels used in the calculations reported herein

The order of orbitals used in all of the calculations reported herein are as illustrated in Figure 1. This has the feature that the maximally entangled orbits, the $1f_{\frac{7}{2}}$ and $2p_{\frac{3}{2}}$, are next to one another as is known to be important based on information theory considerations [9].

4. Calculations for the ground states of even-even nuclei

We first present results for the various even-even nuclei that we have studied, ^{48}Cr, ^{50}Cr, ^{52}Fe, ^{54}Fe and ^{56}Ni, focusing for now on their ground states only. In Table I, we compare the results of J-DMRG calculations with the exact ground state energies as a function of the parameter m that defines how many states are retained in each block.

For ^{48}Cr, we observe smooth convergence to the exact ground state energy as we increase the number of states m retained. However, to achieve this level of accuracy we must retain roughly a quarter of the total number of 41,355 $J^{\pi} = 0^{+}$ shell-model configurations for the system. Even to get reproduction of the exact results to within about 60 keV requires of order 20% of the full space.

^{50}Cr, with two additional neutrons, displays similar features, likewise converging smoothly to the exact result with m and likewise requiring almost 20 % of the full space to obtain agreement to within roughly 60 keV.

Next we consider ^{52}Fe, for which the full dimensionalities are much larger. Here the total number of 0^{+} configurations is 1,777,116. Once again the J-DMRG procedure converges smoothly to the exact result with increasing m. Here a calculation with $m = 200$, containing roughly 4% of the full space, was able to achieve agreement with the exact ground state energy to within 60 keV. ^{52}Fe is the largest even-even nucleus that can be solved exactly with the *public version* of the shell-model code ANTOINE.

Next we turn to ^{54}Fe which has an even larger space, containing 5,220,621 $J^{\pi} = 0^{+}$ states. Here too we see monotonic convergence with m. Furthermore, we find that to be able to achieve accuracy to within roughly 60 keV we now need roughly 3% of the full space.

Finally, we turn to ^{56}Ni, which had been treated earlier, albeit with the KB3 hamiltonian. Here too we see smooth and rapid convergence to the exact results, with substantially less than 1% of the space required to achieve agreement to within 60 keV.

In Figure 2, we summarize results for the even-even nuclei ^{48}Cr, ^{50}Cr, ^{52}Fe and ^{56}Ni, focussing on the number of states needed to obtain 60 keV accuracy as a function of the logarithm of the size of the shell-model basis. We also include a quadratic fit to the results. By extrapolating this fit, we conclude that the method can be used to treat systems with well in excess of 10^{9} 0^{+} configurations, using DMRG matrices of at most a few hundred thousand states.

Journal of Physics: Conference Series **322** (2011) 012002 doi:10.1088/1742-6596/322/1/012002

Table 1. Calculated ground-state energies in MeV as a function of m for several even-even nuclei in the $1f-2p$ shell. The maximum dimensions encountered in the sweep process are also given.

^{48}Cr	m	E_{GS}	Max Dim
	100	-99.475	4,554
	120	-99.495	6,361
	140	-99.514	8,377
	160	-99.560	9,996
	180	-99.573	10,906
	Exact	-99.578	41,355
^{50}Cr	m	E_{GS}	Max Dim
	100	-122.807	34,454
	120	-122.814	42,669
	140	-122.821	51,408
	160	-122.839	62,440
	180	-122.850	73,830
	200	-122.873	83,393
	Exact	-122.877	267,054
^{52}Fe	m	E_{GS}	Max Dim
	120	-151.945	33,377
	140	-151.948	39,501
	160	-151.988	50,578
	180	-152.012	62,277
	200	-152.069	72,317
	Exact	-152.129	1,777,116
^{54}Fe	m	E_{GS}	Max Dim
	140	-175.653	123,639
	160	-175.656	141,397
	180	-175.665	168,026
	200	-175.680	198,253
	Exact	-175.730	5,220,621
^{56}Ni	m	E_{GS}	Max Dim
	100	-205.643	87,633
	120	-205.651	106,383
	140	-205.652	123,196
	160	-205.659	139,966
	180	-205.661	166,695
	200	-205.670	199,274
	Exact	-205.709	15,443,684

The figure does not include our results for ^{54}Fe, which as noted earlier we also treated but which lies somewhat above the curve. We have concluded that this is most likely a reflection of the difference in the number of active protons and neutrons in this case. We have found, however, that this can be improved substantially by choosing different m values for neutron and protons, scaled in proportional to their respective number of (nJ) *partitions*. In particular, if we assume the same number of partitions for both, we find that the fraction required for 60 keV accuracy is 0.031. When we scale m_n and m_p in accord with the relative number of neutron and proton partitions, respectively, the fraction goes down to 0.021. Such a procedure, which is now fully implemented in our J-DMRG algorithm, can be used for any problem with a different number of active neutrons and protons. Indeed, it is possible to improve our treatment even further through the use of a fully dynamical procedure for choosing m values in a sweep [3].

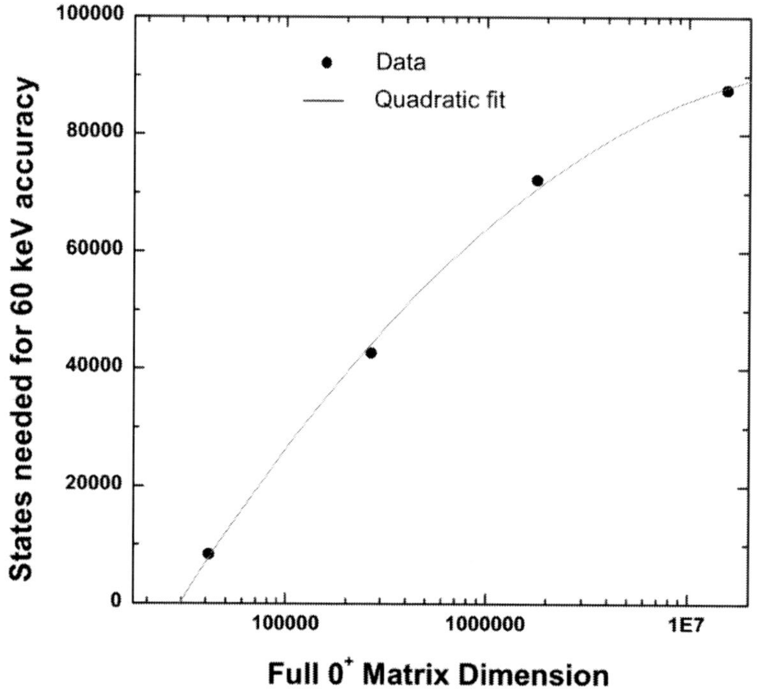

Figure 2. Number of states required for 60 keV accuracy in DMRG calculations of even-even nuclei

5. Calculations for the ground states of odd-mass nuclei

In Table 2, we present the corresponding ground-state results for the odd-mass nuclei that we have studied, ^{47}Cr, ^{49}Cr, ^{51}Fe, ^{53}Fe and ^{55}Ni.

For ^{47}Cr, the results converge smoothly as a function of m and we can achieve accuracy to within 25 keV. However, to achieve this level of agreement, we require inclusion of over 40% of the full space. In ^{49}Cr, the convergence is again smooth and we achieve reproduction of the exact

Table 2. Calculated ground-state energies in MeV as a function of m for some odd-mass nuclei in the $1f-2p$ shell. The maximum dimensions encountered in the sweep process are also given. The exact results for ^{55}Ni were extracted from [10].

^{47}Cr	m	E_{GS}	Max Dim
	30	-83.722	12,229
	40	-83.735	12,567
	50	-83.749	13,455
	60	-83.754	15,914
	70	-83.757	18,578
	Exact	-83.780	41,498
^{49}Cr	m	E_{GS}	Max Dim
	60	-110.099	56,305
	80	-110.103	71,268
	100	-110.113	90,859
	120	-110.119	110,785
	Exact	-110.169	595,314
^{51}Fe	m	E_{GS}	Max Dim
	60	-136.286	161,315
	80	-136.297	230,126
	100	-136.306	217,328
	Exact	-136.409	3,998,059
^{53}Fe	m	E_{GS}	Max Dim
	50	-162.484	307,968
	60	-162.528	319,970
	70	-162.537	333,996
	Exact	*	21,131,892
^{53}Ni	m	E_{GS}	Max Dim
	50	-189.412	450,916
	70	-189.426	529,911
	90	-189.435	663,686
	Exact	-189.534	63,268,915

result to within 50 keV using roughly 20% of the full space. In comparison to their even-even neighbors, ^{48}Cr and ^{50}Cr, the convergence is slightly slower. For ^{51}Fe and ^{53}Fe, convergence is still smooth and the fraction of the full full space required to achieve accurate reproduction of the exact energies goes down rapidly.

^{55}Ni is the largest odd-mass calculation we report here, with the full space having a dimension of 63,268,915 states. While the results indeed converge smoothly, we are unable to achieve agreement to better than 100 keV in this case.

Note that in all of these nuclei, there are slightly different number of active neutrons and protons. The above results, which are based on equal m values for both neutrons and protons, could be improved slightly by scaling them according to the relative numbers of partitions.

As for even-even systems, the fraction of the space required for a fairly high level of accuracy is seen to go down rapidly with the size of the problem. Though the results are not as striking as for even-even nuclei, they are still suggestive that we can go to somewhat larger odd-mass systems with the J-DMRG method.

6. Excited states

Once an optimal block structure has been determined, even if from properties of the ground state only, we can still use it to estimate excited states. To do this we simply couple the optimal blocks together to the desired angular momentum and then diagonalize the resulting hamiltonian.

We illustrate the results that would emerge from such a prescription for the lowest two 0^+ states in ^{56}Ni in Table 3. We also include for comparison the exact energies. What we see is that not surprisingly the description of the ground state is better than the description of the first excited 0^+ state.

It is possible, however, to introduce a mixed density matrix in which we find the optimal block structures required to simultaneously describe more than one state. This at least in principle should enable us to get a comparable description of several states at the same time. In Table 4, we show results again for the two lowest 0^+ states in ^{56}Ni, this time targeting both states *equally* in building the requisite reduced density matrices. Now the description of the second 0^+ state improves dramatically (by roughly 850 keV), without a significant loss of accuracy for the ground state.

Using the mixed density matrices that emerge from this and other calculations, we have evaluated other low-lying excited states in the various even-even nuclei we have treated, but we do not show those results here. In a more elaborate scheme, different weights, determined a priori or dynamically, could be associated with the different excited states to construct the

Table 3. Calculated energies for the ground state and the first excited 0^+ state in ^{56}Ni in MeV, for selected m values. The calculations targeted the ground state alone to build the optimal block structures. The exact results are extracted from [11].

m	$E_{0_1^+}$	$E_{0_2^+}$
140	-205.652	-200.913
160	-205.659	-200.902
Exact	-205.709	-202.092

Table 4. Calculated energies for the ground state and the first excited 0^+ state in ^{56}Ni in MeV, for selected m values. The calculations targeted both the ground state and the first excited 0^+ state to build the optimal block structures. The exact results are from [11].

m	$E_{0_1^+}$	$E_{0_2^+}$
140	-205.641	-201.739
160	-205.645	-201.752
Exact	-205.709	-202.092

Table 5. Calculated $7/2_1^-$ and $3/2_1^-$ energies of ^{53}Fe in MeV as a function of m . In the first set of results only the $7/2_1^-$ ground state was targeted. In the second set of results both the ground state and the first excited state were targeted.

	Target $7/2_1^-$		Target $7/2_1^-$ and $3/2_1^-$	
m	$E_{7/2_1^-}$	$E_{3/2_1^-}$	$E_{7/2_1^-}$	$E_{3/2_1^-}$
50	-162.528	-161.546	-162.484	-161.735
60	-162.530	-161.606	-162.528	-161.743

mixed density matrix.

There is a useful feature that arises when treating odd-mass systems as compared to even-even systems. In even-even systems, there is a large disparity between the dimensions that arise in calculations of the ground state and most low-lying excited states, notably those that do not have spin and parity 0^+. It is for that reason that we targeted the ground state and the first excited 0^+ state of ^{56}Ni when we built mixed density matrices to improve our excited-state description. In odd-mass nuclei, the ground state and other low-lying excitations typically have similar dimensions and thus can be targeted simultaneously with no great computational burden. As such, we have been able to use mixed density matrices more effectively for odd-mass nuclei than for their even-even neighbors [3]. As an example, we illustrate in Table 5 our results for both the ground state and the first excited state of ^{53}Fe. The first set of numbers arise when we target the $7/2^-$ ground state only in our truncation algorithm. The second set derives when we simultaneously target both the ground state and the first excited $3/2^-$ state. Targeting both lowers the $3/2^-$ state by almost 140 keV with essentially no impact on the energy of the ground state.

7. Summary and outlook

In this talk, we have summarized some of our most recent results obtained using the J-DMRG method for nuclei in the $1f - 2p$ shell. We reported systematic calculations of both even-even and odd-mass nuclei for nuclei ranging from ^{47}Cr through ^{56}Ni. Where possible, we compared with the results of exact diagonalization to assess the accuracy and applicability of the method. All calculations converged smoothly to the exact shell-model results. Furthermore, we found that the method scales very well with the size of the problem, with the fraction of the complete space required for a quantitative reproduction of the exact results going down rapidly with the full dimensionality. Though especially true for even-even systems, it was also found to be the case for odd-mass systems, albeit to a lesser extent. The fact that the fraction of the space goes down with the size of the problem bodes well for the future usefulness of the method in even larger shell-model problems.

We are also showed some first results to assess the usefulness of targeting more than just the ground state in the density matrix truncation algorithm. We found that such a method is particularly useful for odd-mass nuclei, where it is possible to simultaneously target several low-lying states with different values of the angular momentum.

There are several issues we are now exploring. One concerns how to treat even larger single-particle orbitals, as arise in heavier nuclei. We have developed a possible procedure for accomplishing this and preliminary application to the $g_{9/2}$ orbital in the Ni isotopes suggests it works well [3]. We are also in the process of including in our formalism the calculation of observables that relate the properties of neighboring nuclei, e.g. beta decay. Finally, we are also more systematically studying the usefulness of targeting more than just the ground state in the density matrix truncation algorithm, especially for odd-mass nuclei.

The primary goal of this work has been to continue testing the J-DMRG method as a potentially practical dynamical truncation strategy for large-scale shell-model calculations of atomic nuclei. The results reported here are quite promising. Once we are able to implement the few improvements suggested above, we expect to be in position for many useful applications of the method to problems of contemporary importance in nuclear structure physics.

Acknowledgments
This work was supported in part by the US National Science Foundation under grant # PHY-0854873. We would like to thank M. Horoi and J.R. Gour for providing us with the exact ^{56}Ni and ^{55}Ni energies, and A. Brown for providing us with the GXPF1A matrix elements. We are also grateful to NERSC for their support of this project through continued allocation of computational resources.

References
[1] White S R 1992 *Phys. Rev. Lett.* **69** 2863
[2] Dukelsky J and Pittel S 2004 *Rep. Prog. Phys.* **67** 513
[3] Bhupender Thakur 2010 *Ph.D. thesis, University of Delaware*
[4] Pittel S and Thakur B 2011 *Acta Physica Polonica B* **42** 427
[5] Pittel S and Sandulescu N 2006 *Phys. Rev. C* **73**, 014301(R)
[6] Thakur B, Pittel S and Sandulescu N 2008 *Phys. Rev. C* **78** 041303
[7] Honma M, Otsuka T, Brown B A and Mizusaki T 2005 *Eur. Phys. J. A* **25** s01 499
[8] Poves A and Zuker A P 1981 *Phys. Rep.* **70** 235
[9] Legeza Ö and Sólyom J 2003 *Phys. Rev. B* **68** 195116
[10] Horoi M, Gour J R, Wloch M, Lodriguito M D, Brown B A and Piecuch P 2007 *Phys. Rev. Lett.* **98** 112501
[11] Horoi M, Brown B A, Otsuka T, Honma M and Mizusaki T 2006 *Phys. Rev. C* **73** 061305(R)

Journal of Physics: Conference Series **322** (2011) 012003 doi:10.1088/1742-6596/322/1/012003

Seniority isomers in nuclei

P Van Isacker

Grand Accélérateur National d'Ions Lourds (GANIL), CEA/DSM–CNRS/IN2P3, Bvd Henri Becquerel, F-14076 Caen, France

Abstract. Seniority isomers are nuclear states with an electromagnetic decay that is hindered by selection rules related to the seniority quantum number. A simple analysis is presented of their possible formation with reference to the nickel isotopes $^{70-76}$Ni and the $N = 50$ isotones from molybdenum to cadmium. It is shown that the existence of seniority isomers in a $j = 9/2$ shell is predominantly governed by the quadrupole pairing matrix element of the nucleon-nucleon interaction. The analysis is generalized to shells with larger j.

Isomers are metastable quantum states. In nuclei isomers generally adopt a configuration which is different from those of states at lower energy and their decay is therefore hindered. The type of configuration change determines the nature of the isomer and hence one distinguishes, for example, shape isomers, spin isomers, and K isomers [1]. Seniority isomers exist by virtue of a conserved quantum number, seniority (denoted as v), and its associated selection rules. Seniority, introduced by Racah in the context of atomic physics [2], refers to the number of particles that are not in pairs coupled to angular momentum $J = 0$. Generally, and in particular in semi-magic nuclei, states with low seniority occur at low energy. For example, the ground state of an even-even semi-magic nucleus has approximately $v \approx 0$ (all nucleons in pairs coupled to $J = 0$) while its yrast levels with angular momenta $J = 2, 4, 6, \ldots$ have $v \approx 2$ (containing one 'broken' pair with $J \neq 0$) [3]. Seniority isomerism is expected to occur in semi-magic nuclei because electric quadrupole (E2) transitions between $v = 2$ states are small when the valence shell is close to half-filled. This result is a consequence of the fact that the matrix elements of even tensor operators—and hence also of the quadrupole operator—between states with the same seniority vanish at mid shell [3].

Examples of seniority isomers have been found in the $N = 50$ isotones with protons dominantly confined to the $\pi g_{9/2}$ shell. In particular, the $J^\pi = 8^+$ levels in ^{94}Ru ($Z = 44$) and ^{96}Pd ($Z = 46$) have half-lives of 71 and 2.2 μs, respectively, resulting from a combination of slow E2 decay and a small energy difference with the $J^\pi = 6^+$ level below it. A review is given by Grawe *et al.* [4]. On the basis of similar arguments one would expect the same phenomenon to occur in the neutron-rich nickel ($Z = 28$) isotopes ^{72}Ni and ^{74}Ni with neutrons dominantly confined to the $\nu g_{9/2}$ shell but this does not seem to be the case [5]. In this contribution a simple explanation of this observation is given. The occurrence of seniority isomers is analyzed in nuclei where valence nucleons of one type (*i.e.*, either neutrons or protons) are confined to a single shell with $j = 9/2$ and this analysis is subsequently generalized to shells with larger j.

The starting point of the discussion is that the seniority classification is a very good approximation for n identical nucleons in a $j = 9/2$ shell (see also chapter 21 of ref. [3]). This is a trivial statement if only one state exists for a given particle number n and angular momentum J, in which case seniority must be exact. If two states with different seniority and the same

Table 1. Coefficients $a^\lambda_{\alpha vJ}$ in the expansion (1) of diagonal energies of the $(9/2)^4$ system.

J^π	v	$\lambda=0$	$\lambda=2$	$\lambda=4$	$\lambda=6$	$\lambda=8$
0^+	0	$\frac{8}{5}$	$\frac{1}{2}$	$\frac{9}{10}$	$\frac{13}{10}$	$\frac{17}{10}$
	4	—	$\frac{13}{66}$	$\frac{735}{286}$	$\frac{961}{330}$	$\frac{459}{1430}$
2^+	2	$\frac{3}{5}$	$\frac{142}{99}$	$\frac{67}{55}$	$\frac{442}{495}$	$\frac{102}{55}$
	4	—	$\frac{134}{99}$	$\frac{241}{143}$	$\frac{842}{495}$	$\frac{901}{715}$
4^+	2	$\frac{3}{5}$	$\frac{67}{99}$	$\frac{746}{715}$	$\frac{1186}{495}$	$\frac{918}{715}$
	4_s	—	$\frac{68}{33}$	1	$\frac{13}{15}$	$\frac{114}{55}$
	4	—	$\frac{47}{99}$	$\frac{300}{143}$	$\frac{1214}{495}$	$\frac{697}{715}$
6^+	2	$\frac{3}{5}$	$\frac{34}{99}$	$\frac{1186}{715}$	$\frac{658}{495}$	$\frac{1479}{715}$
	4_s	—	$\frac{19}{11}$	$\frac{12}{13}$	1	$\frac{336}{143}$
	4	—	$\frac{61}{198}$	$\frac{545}{286}$	$\frac{479}{198}$	$\frac{391}{286}$
8^+	2	$\frac{3}{5}$	$\frac{6}{11}$	$\frac{486}{715}$	$\frac{87}{55}$	$\frac{1854}{715}$
	4	—	$\frac{9}{22}$	$\frac{405}{286}$	$\frac{45}{22}$	$\frac{609}{286}$

Table 2. Coefficients b_J in the expression (2) of the off-diagonal matrix elements of the $(9/2)^4$ system.

J^π	0^+	2^+	4^+	6^+	8^+
b_J	$-\frac{1}{10\sqrt{429}}$	$\frac{\sqrt{14}}{495\sqrt{13}}$	$-\frac{\sqrt{314}}{6435}$	$\frac{\sqrt{281}}{6435}$	$-\frac{\sqrt{19}}{2145}$

J occur, it can be shown that for any reasonable nuclear interaction $\hat{V}$ the off-diagonal matrix element is small compared to the states' energy difference. The argument can be illustrated with the example of the four-particle system $(9/2)^4$ for which the diagonal energies are

$$E[j^4\alpha vJ] = \sum_\lambda a^\lambda_{\alpha vJ}\,\nu_\lambda, \tag{1}$$

while the off-diagonal matrix elements are

$$b_J(65\nu_2 - 315\nu_4 + 403\nu_6 - 153\nu_8), \tag{2}$$

with the coefficients $a^\lambda_{\alpha vJ}$ and b_J listed in tables 1 and 2. In these expressions $\nu_\lambda \equiv \langle j^2\lambda|\hat{V}|j^2\lambda\rangle$ are the two-body interaction matrix elements and α is a quantum number additional to v and J. If we consider as an example the two $J^\pi = 2^+$ states, we find that the 'typical' splitting between the seniority $v = 2$ and $v = 4$ levels is about 1 MeV while the off-diagonal matrix element (2) is only of the order of a few tens of keV.

The proof that seniority mixing is negligible for the $J^\pi = 4^+$ and 6^+ states of a $(9/2)^4$ system is more subtle and is crucial for the existence of seniority isomers. There are *three* states for each of these angular momenta, two of which, with $v = 2$ and $v = 4$, are close in energy and could possibly strongly mix. The $v = 4$ members of these closely-spaced doublets are, however, the so-called 'solvable' $J^\pi = 4^+$ and 6^+ states discussed in refs. [6, 7, 8] which *have exact seniority*

 doi:10.1088/1742-6596/322/1/012003

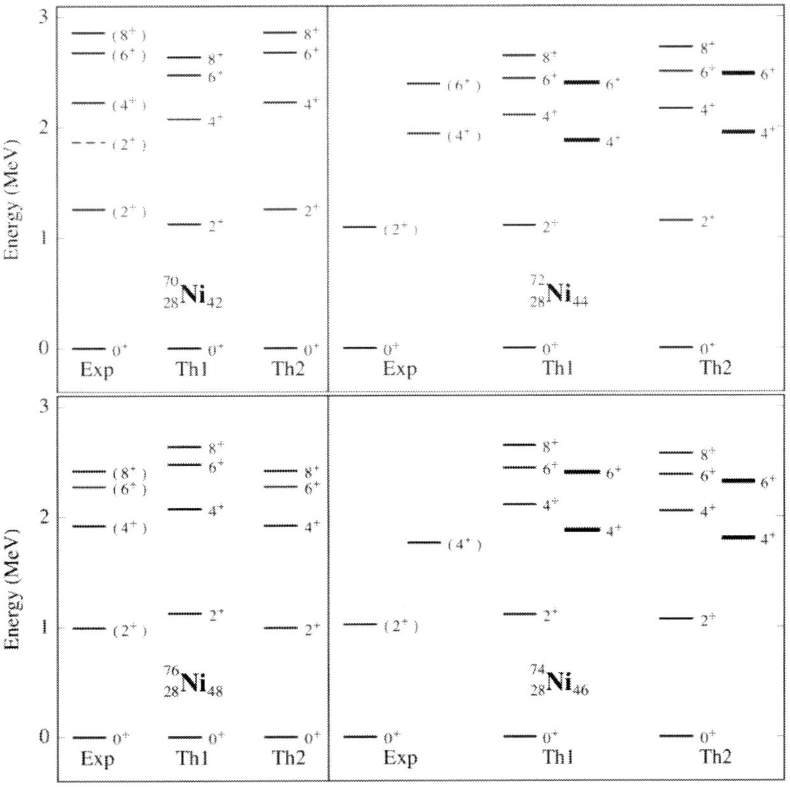

Figure 1. The low-energy spectra of the nickel isotopes $^{70-76}$Ni. The left columns show the observed levels while the columns 'Th1' and 'Th2' contain the results of a $(\nu g_{9/2})^n$ shell-model calculation with constant or linearly varying two-body matrix elements, respectively. The two solvable $J^\pi = 4^+$ and $J^\pi = 6^+$ levels with seniority $v = 4$ are shown with thick lines; the dashed line corresponds to an intruder level.

$v = 4$ *for any interaction* (denoted as $v = 4_s$ in table 1). As a consequence, breaking of seniority only arises through the mixing matrix elements (2) between the $v = 2$ and the higher-lying $v = 4$ states and, by the same argument as above, this mixing is found to be small.

The spectra of the nickel isotopes from ^{70}Ni to ^{76}Ni are compared in fig. 1 to the observed levels [9, 10]. The nucleus ^{70}Ni should, in the approximation of an isolated $\nu g_{9/2}$ shell, display a two-neutron-particle spectrum $(\nu g_{9/2})^2$ with excited states with $J^\pi = 2^+, 4^+, 6^+$ and 8^+. This is indeed found to be the case except for the additional (intruder) $J^\pi = (2^+)$ level at 1867 keV. The nucleus ^{76}Ni displays a two-neutron-hole spectrum $(\nu g_{9/2})^{-2}$ with the same yrast sequence as in ^{70}Ni but no known intruder state. The two-particle and the two-hole spectra fix the differences $\nu_{\lambda-0} \equiv \nu_\lambda - \nu_0$. This is done at two levels of sophistication by taking either constant matrix elements that are the average of those in ^{70}Ni and ^{76}Ni (Th1) or by letting them vary linearly from ^{70}Ni to ^{76}Ni (Th2). In the latter approximation the spectra of the two-particle and the two-hole nuclei are exactly reproduced [except for the intruder $J^\pi = (2^+)$ state in ^{70}Ni]; the description of the two intermediate isotopes, ^{72}Ni and ^{74}Ni, should be rather accurate, albeit very phenomenological.

The two solvable $J^\pi = 4^+$ and $J^\pi = 6^+$ levels with seniority $v = 4$ are shown with thick lines. A noteworthy feature of the calculated spectra of the isotopes ^{72,74}Ni is the occurrence of *two* levels for $J^\pi = 4^+$ and for $J^\pi = 6^+$ which are very close in energy, especially in the latter case.

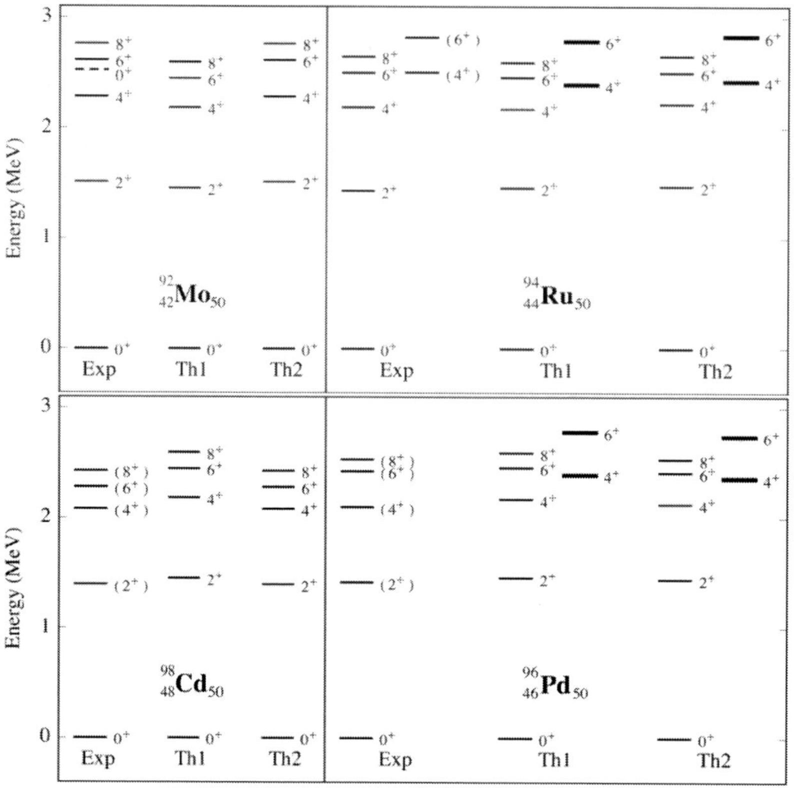

Figure 2. The low-energy spectra of the $N = 50$ isotones ^{92}Mo, ^{94}Ru, ^{96}Pd and ^{98}Cd. The left columns show the observed levels while the columns 'Th1' and 'Th2' contain the results of a $(\pi g_{9/2})^n$ shell-model calculation with constant or linearly varying two-body matrix elements, respectively. The two solvable $J^\pi = 4^+$ and $J^\pi = 6^+$ levels with seniority $v = 4$ are shown with thick lines; the dashed line corresponds to an intruder level.

This is a direct consequence of the solvability of one member of each doublet which cannot mix with the close-lying, predominantly $v = 2$ state with the same spin. At least for the $J^\pi = 6^+$ levels, this feature is still clearly present in the large-scale shell-model calculations of Lisetskiy *et al.* [11].

A similar analysis in the same approximation can be performed for the $N = 50$ isotones with protons in the $\pi g_{9/2}$ shell. In fig. 2 the results of the calculation are compared with the observed spectra [12, 13, 14, 15]. In ^{94}Ru (4_2^+) and (6_2^+) levels are observed at energies of 2503 and 2818 keV, respectively [13]; these are possible candidates for the solvable states which are calculated at 2422 and 2828 keV in the approximation 'Th2'. It would be of interest to confirm the spin assignment and to determine the E2 transition probability between the two levels.

There is a striking difference between the calculated four-particle (or four-hole) spectra of the nickel isotopes and those of the $N = 50$ isotones: the solvable $J^\pi = 4^+$ and $J^\pi = 6^+$ states are yrast in ^{72}Ni and ^{74}Ni while they are yrare in ^{94}Ru and ^{96}Pd. For this reason one may conjecture that the observed yrast (4^+) and (6^+) levels in ^{72}Ni and the observed yrast (4^+) level in ^{74}Ni are the solvable states in question, as is done in fig. 1. This drastically changes the E2-decay pattern as is illustrated in fig. 3. On the left-hand side is shown the 'typical' decay with small $B(\mathrm{E2})$ values between $v \approx 2$ states which is characteristic of the seniority classification in nuclei near mid-shell ($n \approx j + 1/2$) and which is at the basis of the explanation of seniority isomers [4].

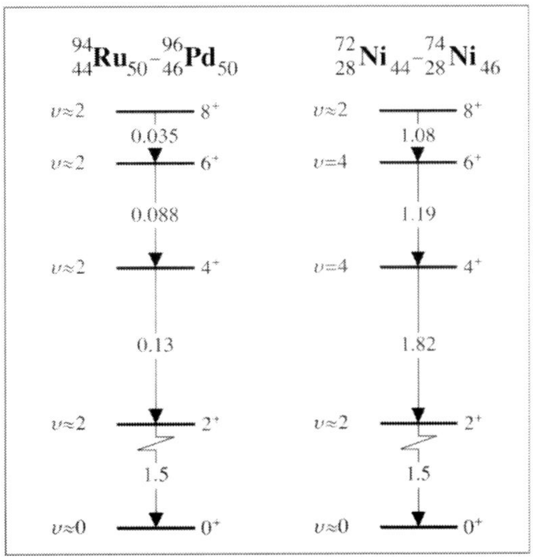

Figure 3. E2 decay between yrast states of the $(9/2)^4$ system as expected in the $N = 50$ isotones (left) and the $Z = 28$ nickel isotopes (right). The numbers between the levels denote calculated $B(\mathrm{E2})$ values expressed in units of $B(\mathrm{E2}; 2_1^+ \to 0_1^+)$ of the two-particle system, assuming a seniority-conserving interaction. All levels have $v \approx 2$, except the $J^\pi = 4^+$ and 6^+ states in ^{72,74}Ni which have exact $v = 4$.

This situation applies to ^{94}Ru and ^{96}Pd where the $v \approx 2$ states are yrast. On the right-hand side of fig. 3 is shown the E2 decay pattern as it is calculated in ^{72}Ni and ^{74}Ni where the yrast $J^\pi = 4^+$ and 6^+ levels have $v = 4$, leading to fast E2 decay in the yrast band. As a consequence, the $J^\pi = 8^+$ and 6^+ levels are unlikely to be isomeric in this case.

Given that approximate analytic expressions are available for the energies of all relevant states, conditions for the existence of isomers can easily be established. Coefficients $a^\lambda_{\alpha v J}$ crucial to the existence of isomers are related to the quadrupole pairing matrix element which has $\lambda = 2$ (see table 1). Although conditions can be derived in general, results become more transparent by averaging over matrix elements with $\lambda \neq 0, 2$ and introducing $\bar\nu \equiv (\nu_{4-0} + \nu_{6-0} + \nu_{8-0})/3$. (This procedure is justified since the three matrix elements are close in value for any reasonable nuclear interaction.) With the help of table 1 one shows that the condition to have yrast $J^\pi = 4^+$ and 6^+ levels with $v = 2$ is

$$\nu_{2-0} \gtrsim \frac{388}{685}\bar\nu \approx 0.57\bar\nu. \tag{3}$$

A second set of conditions for the existence of seniority isomers is obtained by requiring that the $v = 4$ state with angular momentum $J - 2$ is higher in energy than the $v = 2$ state with angular momentum J. This leads to the conditions

$$\nu_{2-0} \gtrsim \frac{553}{850}\bar\nu \approx 0.65\bar\nu, \qquad \nu_{2-0} \gtrsim \frac{32}{65}\bar\nu \approx 0.49\bar\nu, \tag{4}$$

necessary for the existence of a $J^\pi = 6^+$ and a $J^\pi = 8^+$ isomer, respectively. Since both conditions (3) and (4) are necessary, we conclude that $\nu_{2-0} \gtrsim 0.57\bar\nu$ suffices for the existence of a $J^\pi = 8^+$ isomer while $\nu_{2-0} \gtrsim 0.65\bar\nu$ is required for the isomeric character of the $J^\pi = 6^+$. Obviously, these are only approximate conditions, derived under several simplifying assumptions, but they are borne out by the observations in the $N = 50$ isotones and the nickel isotopes, that is, a $J^\pi = 8^+$ isomer exists in the former [4] and so far none is observed in the latter [5].

Table 3. Coefficients $a^{\lambda}_{\alpha v J}$ (with $\alpha = [II']$) in the expansion (1) of the diagonal energies of the $(11/2)^4$ system.

J^π	v	$[II']$	$\lambda=0$	$\lambda=2$	$\lambda=4$	$\lambda=6$	$\lambda=8$	$\lambda=10$
0^+	0	[00]	1.667	0.333	0.600	0.867	1.133	1.400
	4	[22]	—	0.408	2.695	0.061	2.234	0.602
2^+	2	[02]	0.667	1.231	0.952	0.803	0.693	1.654
	4	[22]	—	1.513	0.984	1.098	1.246	1.158
4^+	2	[04]	0.667	0.529	0.926	0.989	1.976	0.913
	4	[22]	—	1.908	0.873	0.792	0.609	1.818
6^+	2	[06]	0.667	0.309	0.684	1.851	1.028	1.461
	4	[24]	—	1.494	1.036	0.717	0.911	1.842
8^+	2	[08]	0.667	0.204	1.046	0.786	1.575	1.722
	4	[26]	—	0.579	1.078	1.082	1.873	1.388
10^+	2	[0, 10]	0.667	0.394	0.391	0.904	1.394	2.249
	4	[28]	—	0.623	0.622	1.340	1.372	2.043

These findings can be generalized to shells with larger j. For $j = 9/2$ the lowest $v = 4$ states with $J^\pi = 4^+$ and 6^+ have a constant structure, independent of the interaction. They can be written as

$$\begin{aligned}
|(9/2)^4, v=4_s, J=4\rangle &= \sqrt{\frac{25500}{25591}}|(9/2)^4[22], v=4, J=4\rangle_{\rm GS} \\
&\quad -\sqrt{\frac{91}{25591}}|(9/2)^4[24], v=4, J=4\rangle_{\rm GS}, \\
|(9/2)^4, v=4_s, J=6\rangle &= \sqrt{\frac{27132}{27257}}|(9/2)^4[24], v=4, J=6\rangle_{\rm GS} \\
&\quad +\sqrt{\frac{125}{27257}}|(9/2)^4[26], v=4, J=6\rangle_{\rm GS}. \qquad (5)
\end{aligned}$$

(The states are identical to those of eq. (8) of ref. [7] but written in an orthonormalized basis.) The labels $[II']$ are intermediate angular momenta and indicate that the state has been obtained by anti-symmetrization of $|j^2(I)j^2(I'); J\rangle$. The states resulting from this anti-symmetrization have been further ortho-normalized by a Gram-Schmidt procedure, as indicated by the subscript 'GS'. The expansion (5) shows that the dominant component of the solvable $J^\pi = 4^+$ state with $v = 4_s$ is $(D^\dagger \times D^\dagger)^{(4)}|o\rangle$, that is, two $J = 2$ pairs coupled to total angular momentum $J = 4$, where it is understood implicitly that this two-pair state is orthogonalized to the $v = 2$ state. For $J^\pi = 6^+$, the dominant component is $(D^\dagger \times G^\dagger)^{(6)}|o\rangle$. This shows that specific two-pair states $[II']$ are a good approximation to the exact $(9/2)^4$ eigenstates and this property can be used to generalize to higher j. The coefficients $a^{\lambda}_{\alpha v J}$ for the two-pair states of relevance in the $(11/2)^4$ system are given in table 3. Such coefficients are rational numbers [16]; since they involve large integers, approximate numerical values are given in the table.

Pure $(11/2)^n$ configurations are found in the $N = 82$ isotones with protons above $Z = 64$ confined to the $\pi h_{11/2}$ shell and taking ^{146}Gd as a closed core [3]. The energy spectrum of the two-particle nucleus ^{148}Dy fixes the two-body interaction. The exact shell-model diagonalization of this interaction for the four-particle nucleus ^{150}Er yields eigenstates that are close to the ones obtained in a two-pair approximation. All exact eigenstates have overlaps of more than 99% with the two-pair states shown in table 3, except for the $J^\pi = 8^+_2$ state which turns out to be a complicated mixture of several two-pair states. Conditions for the existence of seniority isomers can be established for the $j = 11/2$ shell in the same spirit as for $j = 9/2$, and one finds

Table 4. $B(\mathrm{E2})$ values (in units $e^2\mathrm{fm}^4$) observed in $N = 82$ isotones compared with $(11/2)^n$ shell-model results that are exact (SM-1) or in a pair approximation (SM-2).

	$B(\mathrm{E2}; 10_1^+ \to 8_1^+)$			$B(\mathrm{E2}; 8_1^+ \to 6_1^+)$		
	Exp	SM-1	SM-2	Exp	SM-1	SM-2
$A = 148$	43 (4)	43^a	43^a	79 (23)	115	115
$A = 150$	11.4 (1.4)	10.6	10.7	~ 38	27	23
$A = 152$	0.96 (0.14)	0	0	—	0	0

[a]Normalized value.

that the $J^\pi = 10^+$ level is always isomeric while 6^+ and 8^+ are isomeric if $\nu_{2-0} \gtrsim 0.58\bar{\nu}$ and $\nu_{2-0} \gtrsim 0.48\bar{\nu}$, respectively, where $\bar{\nu} \equiv (\nu_{4-0} + \nu_{6-0} + \nu_{8-0} + \nu_{10-0})/4$. The isomeric character of the $J^\pi = 10^+$ and 8^+ levels is established experimentally [17] in ^{148}Dy, ^{150}Er and ^{152}Yb (only 10^+ in the latter), and the resulting $B(\mathrm{E2})$ values are summarized in table 4. The pair approximation (SM-2) is seen to be close to the exact shell-model calculation (SM-1) which in turn is close to the observed E2 decays. It would be of interest to search for isomers in the heavier $N = 82$ isotones ^{154}Hf and ^{156}W. Given that this analysis *presupposes* an isolated $\pi h_{11/2}$ shell, the detection of seniority isomers would be an indication of the doubly magic nature of the nucleus ^{158}Os, akin to ^{146}Gd.

In summary, a simple analysis has been presented of the possible formation of seniority isomers in nuclei, which shows that their existence is predominantly governed by the quadrupole pairing matrix element of the nucleon-nucleon interaction.

Acknowledgments

I wish to thank my collaborators in this work, Stefan Heinze and Larry Zamick. This work was partially supported by the Agence Nationale de Recherche, France, under contract nr ANR-07-BLAN-0256-03.

References

[1] Walker P and Dracoulis G 1999 *Nature* **399** 35
[2] Racah G 1943 *Phys. Rev.* **63** 367
[3] Talmi I 1993 *Simple Models of Complex Nuclei. The Shell Model and Interacting Boson Model* (Chur, Switzerland: Harwood, Academic).
[4] Grawe H, Górska M, Lipoglavšek M, Nyberg J, Grzywacz R, Lewitowicz M, Rykaczewski K, Maier K H and Schubart R 1997 *Prog. Part. Nucl. Phys.* **38** 15
[5] Sawicka M *et al.* 2003 *Phys. Rev. C* **68** 044304
[6] Escuderos A and Zamick L 2006 *Phys. Rev. C* **73** 044302
[7] Van Isacker P and Heinze S 2008 *Phys. Rev. Lett.* **100** 052501
[8] Qi C 2011 *Phys. Rev. C* **83** 014307
[9] Tuli J K 2004 *Nucl. Data Sheets* **103** 389
[10] Mazzocchi C *et al.* 2005 *Phys. Lett. B* **622** 45
[11] Lisetskiy A F, Brown B A, Horoi M and Grawe H 2004 *Phys. Rev. C* **70** 044314
[12] Baglin C M 2000 *Nucl. Data Sheets* **91** 423
[13] Mills W J, Ressler J J, Austin R A E, Chakrawarthy R S, Cross D S, Heinz A, McCutchan E A and Strange M D 2007 *Phys. Rev. C* **75** 047302
[14] Abriola D and Sonzogni A A 2008 *Nucl. Data Sheets* **109** 2501
[15] Singh B and Hu Z 2003 *Nucl. Data Sheets* **98** 335
[16] Talmi I 2010 *Nucl. Phys. A* **846** 31
[17] http://www.nndc.bnl.gov, National Nuclear Data Center.

On the measurement of quadrupole moments of radioactive nuclei

E Padilla-Rodal

Instituto de Ciencias Nucleares, Universidad Nacional Autónoma de México, Apartado Postal 70-543, México, D.F., 04510, MEXICO

E-mail: padilla@nucleares.unam.mx

Abstract. Electric quadrupole moments provide a direct insight on the single-particle structure or the collective nature of a nuclear state. This article presents a short review on some of the experimental methods available to measure quadrupole moments with emphasis on the reorientation technique in Coulomb excitation and its use in combination with radioactive ion beams.

1. Introduction

The description of a nucleus is a hard to accomplish task. We are dealing with a finite many-body system, formed by fermions, governed by quantum-mechanics and interacting via a complicated force. Ideally, one would like to determine the wave function for each energy eigenstate of this system. However, apart from nuclear properties which are very intuitive such as its mass, radius, and nucleon separation energies, the experimental information that one usually gets from the nucleus, consist on the measurement of expectation values of one- and two-body operators. The challenge is then; try to model the complex dynamics of a nucleus so to reproduce the experimental results in a consistent way. The electric quadrupole moment of the first 2^+ state in even-even nuclei represents a precise comparison point for testing the validity of nuclear models. The combination of well established experimental techniques with new technologies in detection systems and important improvements in the radioactive ion beam production, have allowed to obtain new experimental information on the electromagnetic matrix elements of exotic nuclei that can help to understand the evolution of the nuclear structure towards nuclei far from stability.

The first experimental information that one would like to know about the nucleus are its energy eigenvalues. Unfortunately there are many theoretical Hamiltonians that are able to correctly reproduce an energy spectrum using interactions that strongly differ. To select among these interactions one requires more sensitive experimental information. In principle, electromagnetic transition probabilities that connect two energy states in a single nucleus constitute a more stringent test for theoretical models. However, in order to gain a better understanding on the single-particle structure or the collective nature of a nuclear state, quantities that depend on the properties of two eigenstates $\langle f|H_{EM}|i\rangle$ are not the best alternative, since they may carry compensatory errors in $|i\rangle$ and $|f\rangle$ but still agree with the experimental value. Quantities that involve the one state that is been investigated, and depend on a known interaction, $\langle f|H_{EM}|f\rangle$, are evidently a more suitable probe in this respect. Examples of these quantities are the magnetic dipole and electric quadrupole moments.

The determination of the electric quadrupole moment of an excited nuclear state is based, directly or indirectly, on the measurement of its interaction with an electric field gradient (EFG). Very different experimental techniques have been developed over the years in order to achieve its measurement (see, for example, [1]-[5]).This paper gives a short overview on some of those techniques with emphasis on their applicability to the study of exotic nuclei. Recently, the availability of Radioactive Ion Beams (RIBs) [6]-[12] has widen up the boundaries in nuclear research. For example, the study of light nuclei that are located near the drip-lines, has resulted in the observation of phenomena that traditional nuclear models are not able to predict [13]-[16], challenging our understanding of the nucleus. The experimental information on radioactive nuclei is providing extremely valuable insight about the relevant interactions that dictate the behavior of the nucleus when reaching the limits of nuclear stability.

2. Nuclear moments

The interaction energy of an external potential $V(r)$ with a nuclear charge distribution $\varrho(r)$ can be written as a product:

$$E_Q = \sum_{\lambda,\mu}(-)^{\mu}\, T_{\mu}^{(\lambda)}\, V_{-\mu}^{(\lambda)}\,, \tag{1}$$

where $T_{\mu}^{(\lambda)}$ are the multipole tensors of $\varrho(r)$ and $V_{\mu}^{(\lambda)}$ are the tensor components of the external field. The multipole tensors $T_{\mu}^{(\lambda)}$ are classified as electric if its parity is $\pi = (-)^{\lambda}$ and as magnetic if $\pi = (-)^{\lambda+1}$. The well defined parity and time-reversal properties of the nuclear states, $|\Psi\rangle$, implies that the diagonal matrix elements $\langle\Psi|T_{\mu}^{(\lambda)}|\Psi\rangle$, vanish unless the total angular momentum of the state $J \leq \lambda/2$ and exists only for $\pi = +1$. Non-vanishing multipoles are: $E0$, electric monopole (a number); $M1$, the magnetic dipole (a vector); $E2$, electric quadrupole (a rank-2 tensor); *etc.* Note that this selection rules do not allow to measure the quadrupole moment of nuclear states having $J = 0$ or $1/2$.

In practice we measure a single number for each moment, which is called spectroscopic quadrupole moment, and is given by the expectation value of the zero component of the rank-2 tensor

$$\begin{aligned} Q_s(J) &= \langle J, m = J|T_0^{(2)}|J, m = J\rangle \\ &= \left[\frac{16\pi}{5}\right]^{1/2} \langle J, J|r^2 Y_{2\,0}(\theta,\phi)|J, J\rangle \\ &= -\left[\frac{16\pi}{5}\right]^{1/2} \left[\frac{J(2J+1)}{(2J+1)(2J+3)(J+1)}\right]^{1/2} M_{JJ}^{(2)} \end{aligned} \tag{2}$$

where, in general for the λ-moment one has

$$\begin{aligned} \langle J_s M_s|i^{\lambda} r^{\lambda} Y_{\lambda\mu}|J_r M_r\rangle &\equiv \langle J_s M_s|\mathcal{M}(E\lambda,\mu)|J_r M_r\rangle \\ &= (2J_s+1)^{-1/2}\, \langle J_r\lambda, M_r\mu|J_s M_s\rangle\, M_{rs}^{(\lambda)} \end{aligned} \tag{3}$$

with the reduced matrix element

$$M_{rs}^{(\lambda)} \equiv \langle J_s||i^{\lambda} r^{\lambda} Y_{\lambda}||J_r\rangle \qquad \text{and} \qquad M_{rs}^{(\lambda)} = (-)^{J_r - J_s - \lambda}\, M_{rs}^{(\lambda)}\,. \tag{4}$$

The expectation value of the four remaining components of $T^{(2)}$ can be obtained using the Wigner-Eckart theorem.

In the case that the nuclear charge distribution has an axial symmetry, the spectroscopic quadrupole moment can be directly related to the intrinsic quadrupole moment Q_0 and to the nuclear deformation by the expression

$$Q_s(J) = \frac{3K^2 - J(J+1)}{(J+1)(2J+3)} Q_0 \tag{5}$$

where K is the projection of the total angular momentum of the nucleus along the deformation axis, and Q_0 is related to the deformation parameter β through the relation

$$Q_0 = \frac{3}{5\pi} ZR^2\beta(1+0.36\beta), \qquad \text{for} \qquad R = R_0\, A^{1/3}. \tag{6}$$

3. Experimental methods

A practical classification of the methods that are available in the laboratory to measure electric quadrupole moments can be obtained by looking at the way the EFG is produced at the nuclear site. For many years, the values of EFG produced by external electrodes in the laboratory were too small to yield a measurable effect when interacting with a charge distribution of nuclear dimensions. As an alternative, the EFG produced by electrons in atomic or molecular environments, or by other charged particles (such as muons or kaons in bound orbits), or produced during the bombardment of a target nucleus with energetic ions were used. This section presents a brief outline of these experimental methods.

3.1. Atomic or molecular environments

In a solid, electric field gradients of up to a few 10^{18} V/cm^2 can be obtained from atomic or molecular charges. The hyperfine interaction between this kind of EFG and the quadrupole moment of an excited nuclear state can be measured from: (i) the magnitude of the hyperfine splitting in atomic transitions (energy differences), (ii) the Mössbauer effect (in case of low-energy gamma transitions $E_\gamma \leq 130$ keV); (iii) the attenuation of the angular distribution of deexcitation gamma rays.

3.1.1. Mössbauer effect When a nucleus is in a solid, fixed within a lattice, there is considerable probability that it can absorb and emit γ-rays without energy loss going into the recoiling nucleus (Mössbauer effect [17], [2]). The recoil energy and Doppler broadening of the γ-ray are characterized by the mass and random velocity of the whole crystal and are therefore completely negligible. This high precision resonant absorption effect can be used to investigate very small changes in the nuclear energy levels that result from the hyperfine interactions between the nucleus and its electronic environment.

The probability of such a recoilless event depends on the energy of the gamma-ray, the temperature and vibrational properties of the solid. In general, the Mössbauer effect is optimized for low-energy gamma-rays and low temperature lattices.

The finite energy spread of the Mössbauer gamma ray can be related to the lifetime of the nuclear excited state. Whereas the interaction of the nuclear quadrupole moment with the EFG gives rise to a splitting of the nuclear energy levels that correspond to different alignments of the quadrupole moment with respect to the principal axis of the EFG.

The main drawback of this kind of measurement is that the electric quadrupole moment is a function of both nuclear and electronic properties combined in such a way that independent quantitative information on both properties cannot be obtained. The determination of the nuclear quadrupole moment should then rely on a calculation of the EFG at the nuclear site.

3.1.2. Perturbed angular correlations The angular correlation of a cascade $I_i \rightarrow I \rightarrow I_f$ is altered as soon as the nucleus in the intermediate level I is subject to a torque, due to the interaction of its electric quadrupole moment with the EFG that surrounds the decaying nucleus [4], [18]. If the quantization axis is chosen to coincide with the direction of emission of the first radiation, the interaction will change the relative angle between these two axis (quantization-emission) inducing transitions among the m-states. The second radiation is therefore emitted from a level with an altered m-population distribution and this will cause an attenuation of the angular correlation.

When the extra-nuclear field is static a semiclassical picture is suitable, the population of the m-states quantized along the direction of the first emission remains constant and its change in time is due to the precession of the nucleus around the symmetry axis. However, when the interaction is time-dependent, the direction of the EFG at the position of each nucleus changes continuously in a random manner, no quantization axis exists for which the population of the m-states remains constant. Consequently all the m-states are equally populated for any choice of the quantization axis and the directional correlation becomes isotropic.

The magnitude of a static perturbation can be measured by the precession frequency:

$$\omega_Q = -\frac{e\,Q\,V_{zz}}{4I\,(2I-1)\,\hbar} \tag{7}$$

As a raw criterion [4], if τ represents the mean life of the intermediate level I, the cascade will be perturbed if $\omega\tau \geq 1$. For time-dependent perturbations one has that the m-states approach to an uniform population exponentially, with a relaxation constant λ, so the criterion to have perturbation in the angular correlation becomes $\lambda\tau \leq 1$.

3.1.3. β-NMR on polarized beams In this method, a polarized beam of radioactive ions is implanted into a crystal, characterized by a well known electric field gradient and oriented along the direction of a static magnetic field [5], [19]-[21]. A variable radio frequency magnetic field, applied perpendicular to the static field, is used to scan around the Larmor frequency, giving rise to equidistant magnetic resonances in the β-decay. The distance between resonances gives a measurement of the product of the nuclear quadrupole moment and the electric field gradient at the nucleus site.

Several techniques are used to produce spin-oriented radioactive ion beams. One of them is the optical pumping using collinear polarized laser light. Another method is the use of nuclear reactions such as: fusion-evaporation, projectile fragmentation, Coulomb excitation and transfer reactions.

Most these experiments determine the interaction energy, *i.e.* the product of the quadrupole moment and the EFG, and therefore to obtain meaningful information about the nuclear quadrupole moment an independent knowledge of the EFG is required.

3.2. Muonic X-rays

Negative muons are usually obtained from the decay of negative pions in flight [22]. To stand a chance of been captured into an atomic orbit, muons must be slowed down to energies ≤ 2 keV so that their velocity becomes lower than the velocity of the valence electrons. The initial distribution of states is not well known, however, a muon orbit of the same size as the electron K-orbit ($n_e = 1$) will have a principal quantum number

$$n_\mu \sim \left(\frac{m_\mu}{m_e}\right)^{1/2} \sim 14\,, \tag{8}$$

As all muonic states are unoccupied, the muon will cascade down from $n_\mu = 14$ towards lower energies, until reaching the most deeply bound states: $1s$ (K-shell) and $2p$ (L-shell). The close proximity of these states (muon K-shell is ~ 207 times closer to the nucleus than an electron K-shell) allow muons to probe the nuclear volume sensitively. The interaction strength of the muon with the electric quadrupole moment of the nucleus varies proportional to r^{-3}, and therefore gets enhanced for low energy states.

The influence of the nuclear electric quadrupole moment in a muonic atom is manifested as an energy splitting of its muonic X-ray lines. To extract the nuclear quadrupole moment from this energy difference one needs to use a model for both, the Hamiltonian that describes the muonic-atom and the nuclear charge distribution [3].

3.3. Reoritentation in Coulomb excitation

Coulomb excitation occurs when a nucleus undergoes transitions to its excited states due exclusively to the time dependent electromagnetic interaction acting between target and projectile as they approach to each other in a collision. This mechanism is extremely useful because its cross section can be accurately related to the electromagnetic properties of the nucleus, such as its reduced electromagnetic transition probabilities and moments [23, 24].

A simplified theoretical evaluation of the Coulomb excitation process can be obtained by assuming that the projectile follows closely a classical hyperbolic trajectory (semi-classical theory). The differential cross section for finding the nucleus in a state f after the collision can be written as the product of the Rutherford differential cross section in the center-of-mass system, $d\sigma_R/d\Omega$, and the excitation probability, $P_{i\to f}$

$$\frac{d\sigma_{i\to f}}{d\Omega} = \frac{d\sigma_R}{d\Omega} P_{i\to f} , \tag{9}$$

This probability can be evaluated using a perturbation approach. To estimate the perturbation order that is suitable in each case it is useful to calculate the parameter $\chi^{(\lambda)}_{i\to f}$, that measures the strength of the interaction with which the state J_i is coupled to state J_f by multipole order λ:

$$\chi^{(\lambda)}_{i\to f} = \frac{(16\pi)^{1/2}(\lambda+1)!}{(2\lambda+1)!!} \frac{Z_2 e}{\hbar(v_i v_f)^{1/2}} \frac{\langle J_f||M(E\lambda)||J_i\rangle}{a^{\lambda}_{if}(2J_i+1)^{1/2}} , \tag{10}$$

where $2a_{if}$ is the symmetrized distance of closest approach in a head-on collision. To first-order in perturbation theory, the transition amplitudes are proportional to $B(E\lambda; J_i \to J_f) = (2J_i+1)^{-1} |\langle J_i||M(E\lambda)||J_f\rangle|^2$, the matrix elements of the electric quadrupole operator between the initial and the final excited state. To second-order, the excitation of the state $|f\rangle$ will also be influenced by the static quadrupole moment of $|f\rangle$, which is proportional to the diagonal element $\langle J_f||M(E\lambda)||J_f\rangle$. The change on the excitation observables introduced by the static quadrupole moment of the excited state f is called *"reorientation effect"* [1, 25].

There are several experimental possibilities for measuring reorientation, that can be divided in: (i) absolute measurements, which require the independent determination of both the $B(E2)$ values and the absolute cross section, and (ii) relative measurements, that consist on the comparison of the excitation probabilities for two different bombarding conditions, such as: different reaction partners, different bombarding energies or different scattering angles. To choose the best conditions for a reorientation experiment one has to consider the way the quadrupole moment affect the measured quantity as well as the accuracy and ease with which such an experiment can be performed.

For the particular case of an even-even nucleus the probability of a $0^+ \to 2^+$ transition can be written as:

$$P_{02} = P^{(11)}_{02} + P^{(12)}_{02} + P^{(22)}_{02} , \tag{11}$$

where the first order term is given by

$$P_{02}^{(11)} = \frac{1}{5}\left[\chi_{02}^{(2)}\right]^2 \sum_{\mu}\left[\mathcal{K}_{2\mu}(\theta,\xi_{02})\right]^2 , \tag{12}$$

the second term in (11) is an interference term

$$P_{02}^{(12)} = \frac{1}{\sqrt{5}}\left[\chi_{02}^{(2)}\right]^2 \chi_{22}^{(2)} \sum_{\mu}\left[\mathcal{K}_{2\mu}(\theta,\xi_{02})\, \mathcal{B}_{2\mu}^{(22)}(\xi_{02},0,\theta)\right]^2 , \tag{13}$$

and term in (11) can be neglected as it becomes important only if the interference between first- and third-order is also important. Functions $\mathcal{K}_{\lambda\mu}$ and $\mathcal{B}_{\lambda\mu}^{(\lambda_1\lambda_2)}$ are defined in equations 15 and 27 of Ref. [1]. In all the previous equations θ is the scattering angle of the projectile in the centre of mass, and ξ_{if} is the adiabaticity parameter that gives the ratio of the collision time $\tau_{\rm coll} = a/v$ relative to the period of the nuclear transition $\tau_{\rm nucl} = \hbar/\Delta E$ and is given by

$$\xi_{if} = \frac{Z_1 Z_2 e^2}{\hbar}\left(\frac{1}{v_i} - \frac{1}{v_i}\right) . \tag{14}$$

In terms of the reorientation coefficient

$$r \equiv \frac{P_{02}^{(12)}}{P_{02}^{(11)}} \approx \frac{A_1}{Z_2}\,\frac{\Delta E}{1 + A_1/A_2}\,\langle 2^+||M(E2)||2^+\rangle\; K(\theta,\xi_{if}) \tag{15}$$

the excitation probability can be written as

$$P_{02} \approx P_{02}^{(11)}(1+r) , \tag{16}$$

where

$$K(\theta,\xi) = 1.135\,\frac{\sum_\mu \mathcal{K}_{2\mu}(\theta,\xi_{02})\,\mathcal{B}_{2\mu}^{(22)}(\xi_{02},0,\theta)}{\xi\,\sum_\mu\left[\mathcal{K}_{2\mu}(\theta,\xi_{02})\right]^2} . \tag{17}$$

It is important to notice that the reorientation coefficient, r, is linear on both the excitation energy ΔE and the mass of the projectile, whereas the probability (16) depends on both the $B(E2)$ and the Q values. From this last observation follows that, a challenge bound to any reorientation experiment is try to single out a small interference term in the presence of a large first-order term, and this requires an understanding of all the processes that may influence the measurement.

One way to get around the strong dependency on the $B(E2)$ value is by measuring two excitation probabilities for different bombarding conditions α and β, and taking the ratio

$$\frac{P_\alpha}{P_\beta} \approx \frac{P_\alpha^{(11)}}{P_\beta^{(11)}}\,\frac{1+r_\alpha}{1+r_\beta} \approx \frac{P_\alpha^{(11)}}{P_\beta^{(11)}}\,(1 + r_\alpha - r_\beta) , \tag{18}$$

that cancels out large uncertainties that may arise form an absolute calibration. The deviation of the experimental ratio P_α/P_β from its calculated value is proportional to $(r_\alpha - r_\beta)$, which is proportional to the quadrupole moment of the excited state.

4. Reorientation experiments with radioactive ion beams

In recent years we have been interested in using the reorientation effect to measure the electric quadrupole moment of the neutron-rich nucleus ^{78}Ge. The last systematic study for the stable $Z = 32$ isotopes were performed by a JAERI group using Multiple Coulomb Excitation [26]. Their results confirm the observation that the values of the diagonal matrix element $\langle 2_1^+ || M(E2) || 2_1^+ \rangle$ evolve from a very small but positive value in ^{70}Ge towards small but negative values in ^{72,74,76}Ge [27]. The question we would like to answer with a measurement in ^{78}Ge is how does the quadrupole moment evolves as we move towards neutron-rich nuclei?

We chose bombarding energies below the Coulomb barrier to guarantee that the observed nuclear excitation was due to a pure Coulomb interaction. The RIBs of ^{78}Ge were produced using the Isotope Separation On-Line (ISOL) technique at the HRIBF Facility in Oak Ridge National Laboratory (ORNL) [6]. To single out the reorientation effect from the presence of the large first-order term we performed a relative measurement using two different target materials: ^{12}C and ^{24}Mg mounted on the same holder.

The Coulomb excitation cross section of the projectile-like nuclei was obtained from the coincidence yield of a de-excitation γ-ray detected in the HPGe segmented array CLARION [28] and a target-like nucleus identified by a large angular coverage charged-particle array, Bare-Ball [29], based on CsI(Tl) crystals coupled to Si-photodiodes. The use of the powerful analysis codes GOSIA and GOSIA2 [30] was extremely important in this analysis, in order to achieve a careful balance a large number of effects that came into play in the measurement. A detailed discussion on the analysis procedure and final results of this analysis will be reported in a forthcoming publication [31].

5. Conclusions

The experimental techniques mentioned here are examples of well established methods to determine electric quadrupole moments of nuclear excited states. In the last years some of these techniques have been combined with new detection systems and applied to the study of exotic nuclei using radioactive ion beams.

Quadrupole moments are a very sensitive probe of the interplay between macroscopic and microscopic effects present in a nuclear wave function and can help to understand changes in the nuclear shell structure when approaching to large values of the proton-to-neutron ratio.

Acknowledgments

This research was supported by CONACyT under the grant 103366 and PAPIIT grant IN-121209.

References

[1] De Boer J and Eichler J 1968 *Adv. Nucl. Phys.* **1** 1
[2] Wagner F E *et al* 1972 *Z. Phys.* **254** 112
[3] Steffen R M 1985 *Hyperfine Interact.* **24-26** 223
[4] Steffen R M and Frauenfelder H 1964 *Perturbed Angular Correlations* ed Karlsson E *et al* (Amsterdam: North-Holland Publ.) p 1
[5] Asahi K and Matsuta K 2001 *Nucl. Phys.* A **693** 63
[6] Beene J R *et al* 2011 *J. Phys. G: Nucl. Part. Phys.* **38** 024002
[7] Huyse M and Raabe R 2011 *J. Phys. G: Nucl. Part. Phys.* **38** 024001
[8] Ball G C *et al* 2011 *J. Phys. G: Nucl. Part. Phys.* **38** 024003
[9] Navin A *et al* 2011 *J. Phys. G: Nucl. Part. Phys.* **38** 024004
[10] Van Duppen P and Riisager K 2011 *J. Phys. G: Nucl. Part. Phys.* **38** 024005
[11] Morrissey D J 2011 *J. Phys.: Conf. Ser.* **267** 012001
[12] Suda T 2011 *J. Phys.: Conf. Ser.* **267** 012008
[13] Riisager K 2006 *The Euroschool Lectures on Physics with Exotic Beams Vol. II (Lecture Notes in Physics vol 700)* ed Al-Khalili J and Roeckl E (Berlin: Springer) p 1

Journal of Physics: Conference Series **322** (2011) 012004 doi:10.1088/1742-6596/322/1/012004

[14] Fridmann J *et al* 2005 *Nature* **435** 972
[15] Schiffer J P *et al* 2004 *Phys. Rev. Lett.* **92** 162501
[16] Otsuka T *et al* 2005 *Phys. Rev. Lett.* **95** 232502
[17] Boyle A J F and Hall H E 1962 *Rep. Prog. Phys.* **25** 441
[18] Shirley D A and Haas H 1972 *Annu. Rev. Phys. Chem.* **23** 385
[19] Ogura M *et al* 2004 *Hyperfine Interact.* **159** 235
[20] Nagae D *et al* 2007 *Proc. 17th Int. Spin Physics Symposium (AIP Conf. Proc. vol 915)* ed Imai K *et al* (New York: AIP) p 853
[21] Shimada K *et al* 2007 *Proc. 17th Int. Spin Physics Symposium (AIP Conf. Proc. vol 915)* ed Imai K *et al* (New York: AIP) p 857.
[22] Wu C S and Wilets L 1969 *Annu. Rev. Nucl. Sci.* **19** 527
[23] Alder K and Winther A 1975 *Electromagnetic Excitation* (Amsterdam: North-Holland)
[24] McGowan F K and Stelson P H 1974 *Nuclear Spectroscopy and Reactions: Part C (Pure and Applied Physics 40-C)* ed Cerny J (New York: Academic Press Inc.) p 3
[25] Häusser O 1974 *Nuclear Spectroscopy and Reactions: Part C (Pure and Applied Physics 40-C)* ed Cerny J (New York: Academic Press Inc.) p 55
[26] Toh Y *et al* 2002 *Int. Nuclear Physics Conf.* ed Norman E p 793
[27] Lecomte R *et al* 1982 *Phys. Rev.* C **25** 1667
[28] ⟨http://www.phy.ornl.gov/hribf/equipment/clarion/⟩
[29] ⟨http://www.phy.ornl.gov/hribf/equipment/hyball/⟩
[30] ⟨http://www-user.pas.rochester.edu/~gosia/mediawiki/index.php/Main_Page⟩
[31] García-Ruiz R F 2011 Master in Science Thesis UNAM

Journal of Physics: Conference Series **322** (2011) 012005 doi:10.1088/1742-6596/322/1/012005

Nuclear and particle physics aspects of Big Bang nucleo-synthesis

Osvaldo Civitarese [1] **and Mercedes Mosquera** [2]

[1] Department of Physics, University of La Plata, Argentina

[2] Department of Astronomy and Geophysics, University of La Plata, Argentina

E-mail: osvaldo.civitarese@fisica.unlp.edu.ar

E-mail: mmosquera@fcaglp.unlp.edu.ar

Abstract. Big Bang Nucleo Synthesis (BBN) is the process which has occurred during the first three minutes of the Universe, that is during the time it needed to cooled down to temperatures of the order of 10^9 K. The standard BBN model depends on one parameter, the baryonic density, and it is one of the proper tools to study the physics of the Universe at its earliest stage. In this paper we shall discuss the rol of BBN as a testing ground for evidences and possible limits which could be determined from cosmological analysis based on the inclusion of sterile neutrinos, the variation of the mass of the Higgs boson and the time variation of fundamental constant.

1. Introduction

Cosmology has indeed evolved to the level of an experimental science (even showing nice error bars in its predictions, something which years ago was considered a mere speculation). Nuclear physics has a hand in cosmological studies, and it also has become an unavoidable partner in determining the limits and scope of the cosmological interpretation of data. Out of the many aspects of nuclear-physics cosmological-oriented studies, we have selected two topics: a)the variation of the mass of the Higgs boson, and b)the inclusion of sterile neutrinos in the electroweak decays, to give examples about the way in which nuclear physics and cosmology talk to each other and set a common frontier of interest.

Several observations can establish limits to the variation of different fundamental constants, such as the atomic clocks and the analysis of the spectra of quasar absorption systems. These astronomical observations suggest a possible variation of the fine structure constant and the electron-to-proton mass ratio [1]. However, other analysis of similar astronomical data gives null variation of the fine structure constant [2].

Big Bang Nucleosynthesis (BBN) is an useful tool to study time variation of fundamental constants such as the fine structure constant, α, the Higgs vacuum expectation value, v, the Planck mass among others. Several theories that attempt to unify the four fundamental interactions, such as super-strings, brane world, and Kaluza-Klein theories, allow fundamental constants to vary within cosmological time scales. In a previous work we have reported on the methods which can be used to calculate it [3, 4]. If the Higgs vacuum expectation value acquires a different value during BBN than the present one, the electron mass, the proton-neutron mass difference, the Fermi constant, and the deuterium binding energy, ϵ_D will be different than the corresponding actual values [5, 6].

Journal of Physics: Conference Series **322** (2011) 012005 doi:10.1088/1742-6596/322/1/012005

Here, we study the effects of a possible variation of the Higgs vacuum expectation value, considering a fixed value of the strong-coupling constant Λ_{QCD}. To perform the calculation of the primordial abundances, we use the linear dependence of ϵ_D with v discussed in [3, 4]. We use observational data of D, ^{4}He and ^{7}Li, to obtain constraints on the variation of the participant fundamental constants. We also perform an analysis of the sensibility of these constraints upon the lithium abundance and upon the dependencies between ϵ_D and v.

2. Bounds from BBN

The dependence of the deuterium binding energy on the Higgs vacuum expectation value is model dependent [5]. We have calculated, for four different nucleon-nucleon potentials (Argonne v_{18} potential, Bonn potential, Nijmegen potential and Reid 93 potential), the proportional constant κ that relates ϵ_D and v

$$\frac{\Delta\epsilon_D}{(\epsilon_D)_0} = \kappa\frac{\Delta v}{v_0}, \tag{1}$$

where $\Delta\epsilon_D = (\epsilon_D)_{BBN} - (\epsilon_D)_0$, and $\Delta v = v_{BBN} - v_0$. The subindexes BBN and 0 indicate the value of the constant at primordial nucleo-synthesis and at the present time, respectively. In Table 1 we present the obtained values of κ. The Argonne v_{18} potential includes an electromagnetic interaction, proportional to the fine structure constant. We have modified this potential to include the variation of the fine structure constant and performed the calculation of ϵ_D. We obtained the relation $\frac{\Delta\epsilon_D}{(\epsilon_D)_0} = -0.0019\frac{\Delta\alpha}{\alpha_0}$.

Table 1. Values of the coefficient κ in the relationship $\frac{\Delta\epsilon_D}{(\epsilon_D)_0} = \kappa\frac{\Delta v}{v_0}$ [3, 4].

Potential	κ
Argonne	-1.23
Bonn	-0.66
Nijmegen	-1.66
Reid	-1.83

In order to calculate the primordial abundances, we have modified the numerical code developed by Kawano [7] for each nucleon-nucleon potential of Table 1. For details see [3].

WMAP data are able to constraint the baryon density $\Omega_B h^2$ (related to the baryon-to-photon ratio η_B) with great accuracy, however there is still some degeneracy between the model parameters, namely: one is dealing with a parametric hypersurface defined by the values of $\Omega_B h^2$, $\Omega_{CDM} h^2$ (dark matter density in units of the critical density), Θ (gives the ratio of the co-moving sound horizon at decoupling to the angular diameter distance to the surface of last scattering), τ (re-ionization optical depth), n_s (scalar spectral index), A_s (amplitude of the density fluctuations). For this reason we have computed the light nuclei abundances for the following cases: i)variation of α and v allowing η_B vary, ii)variation of α and v keeping η_B fixed at WMAP value $\left(\eta_B^{WMAP} = (6.108 \pm 0.219) \times 10^{-10}\right)$. In order to obtain the best fit values for the parameters, we have performed a χ^2-test to compare the theoretical abundances and the observational data. We used the latest data of [8] for D, of [9] for ^{4}He and, for ^{7}Li we considered the data given by [10], and all available older data. Regarding the consistency of them, we have followed the treatment of [11], and increased the errors by a fixed factor: $\Theta_{\rm D} = 2.37$, $\Theta_{^4\rm He} = 2.69$ and $\Theta_{^7\rm Li} = 1.43$ for D, ^{4}He and ^{7}Li, respectively.

2.1. Variation of α and v allowing η_B to vary

We have computed the BBN abundances for different values of the fine structure constant, the Higgs vacuum expectation value and η_B, for each potential considered (see Table 1). We performed a χ^2-test in order to find the best-fit-values of the parameters. In Table 2 we present the results.

Table 2. Best fit parameter values at 1σ standard deviations, for the BBN constraints on η_B (in units of 10^{-10}), $\frac{\Delta\alpha}{\alpha_0}$ (in units of 10^{-3}) and $\frac{\Delta v}{v_0}$ (in units of 10^{-3}), for different values of κ. In all cases, the value of $\frac{\chi^2_{min}}{N-3}$ is 1.00.

κ	$\eta_B \pm \sigma\ [10^{-10}]$	$\frac{\Delta\alpha}{\alpha_0} \pm \sigma\ [10^{-3}]$	$\frac{\Delta v}{v_0} \pm \sigma\ [10^{-3}]$
-1.23	$6.440^{+0.382}_{-0.219}$	-2.5 ± 4.8	$29.5^{+1.3}_{-1.1}$
-0.66	$6.150^{+0.365}_{-0.209}$	$-8.5^{+4.5}_{-4.7}$	$29.9^{+1.3}_{-1.1}$
-1.66	$6.744^{+0.317}_{-0.378}$	$1.5^{+4.7}_{-4.0}$	$29.6^{+1.1}_{-1.3}$
-1.83	$6.901^{+0.243}_{-0.386}$	3.0 ± 4.0	$29.6^{+1.0}_{-1.4}$

The results for all potentials are similar, except for the sign of the relative variation of α. We found good agreement, at the level of three standard deviations, between our best-fit-value of η_B and the one obtained using WMAP data [12]. We also found null variation of α within two standard deviations, and variation of the Higgs vacuum expectation value, at the level of six standard deviations.

2.2. Variation of α and v keeping η_B fixed

Once again, we have computed the BBN abundances for different values of the fine structure constant and of the Higgs vacuum expectation value, for each potential considered (see Table 1). This calculation have been made keeping η_B fixed at WMAP value $\left(\eta_B^{WMAP} = (6.108 \pm 0.219) \times 10^{-10}\right)$ [12]. We have performed a χ^2-test in order to find the best-fit-value. In Table 3 we present the results.

Table 3. Best fit parameter values, within one standard deviation, (1σ errors), for the BBN constraints on $\frac{\Delta\alpha}{\alpha_0}$ (in units of 10^{-3}) and $\frac{\Delta v}{v_0}$ (in units of 10^{-3}), with η_B fixed at the WMAP estimation, for the different values of κ.

κ	$\frac{\Delta\alpha}{\alpha_0} \pm \sigma\ [10^{-3}]$	$\frac{\Delta v}{v_0} \pm \sigma\ [10^{-3}]$	$\frac{\chi^2_{min}}{N-2}$
-1.23	-2.0 ± 4.0	29.0 ± 1.5	1.04
-0.66	-9.0 ± 5.0	29.5 ± 2.0	0.96
-1.66	3.0 ± 4.0	28.5 ± 1.5	1.19
-1.83	5.0 ± 4.0	28.0 ± 2.0	1.25

We found null variation of α at 2σ level, however, the variation of v is not-null even at 6σ. If the analysis does not include the lithium data set, we found null variation in both fundamental constants at the level of 2σ.

In summary, we have obtained bounds on the joint variation of α and v using the observational abundances of D, ^{4}He and ^{7}Li, and performed the analysis for different estimations of the dependence of the deuterium binding energy on the Higgs vacuum expectation value. We have found that the four dependencies give similar results, leading to reasonable fits for the variation of α, v and η_B for the whole data set. We only found variation of v when the ^{7}Li abundance is included in the statistical analysis. If the present values of ^{7}Li abundances are correct, varying fundamental constants would be a possible candidate for solving the discrepancy between the light elements abundances and the WMAP estimates.

3. Sterile neutrinos

In recent papers [13, 14] the sensitivity of the ^{4}He primordial abundance, upon distortions of the light neutrino spectrum induced by couplings with a sterile neutrino, was analyzed. The effects due to the mixing between sterile and active neutrinos reflect upon Big Bang Nucleosynthesis (BBN) in a noticeable manner. Previous studies on this matter can be found in [15] and [16]. The results of [13] may be taken as a solid starting point for a systematic analysis of the sterile-active neutrino mixing upon cosmological observables. By the other hand, the mixing mechanism between sterile and active neutrinos has been studied in detail (see [17] and references therein), so that the calculation of neutrino distribution functions can readily be performed. The information about the neutrino distribution function, in the flavor basis and at a given temperature, is an essential element in the calculation of the neutron decay rate, which is a critical quantity entering BBN. In the framework of the standard cosmological model, sterile neutrinos would produce a faster expansion rate for the Universe and a higher yield of ^{4}He. This is, indeed, a severe constraint on neutrino mixing since a higher predicted abundance of ^{4}He may be in conflict with observational data [19]. Another constraint on active-sterile neutrino mixing is the neutrino mass derived from Cosmic Microwave Background Anisotropy (CMB) [21].The analysis of constraints presented in [22] focus on the mixing scheme at the level of the neutrino mass hierarchy, and it suggests the adequacy of the non-degenerate mass hierarchy to set limits on the mass difference between active and sterile neutrinos, δm^2_{a-s}.

In standard BBN calculations, the mixing of sterile and active neutrinos affects the leptonic fractional occupancies, which are essential quantities appearing in the expression of the weak decay rates. Thus, one needs to know, as input of the calculations, the parameters of the proposed mixing scheme, the neutrino mass hierarchy and the leptonic densities [20]. With these elements one can calculate neutron-decay-rates and neutron abundances, by assuming the freeze-out of weak interactions. The effective number of neutrino generations, N_ν, is fixed by the analysis of CMB . Current limits on the neutrino degeneracy parameter, for light (electron) neutrinos, η_l [14], runs from -0.1 to 0.3.

Here we focus on the calculation of the abundances of D, ^{3}He, ^{4}He and ^{7}Li, in presence of sterile-active neutrino mixing in the three flavor scenario, and for the normal and inverse neutrino mass hierarchies [17]. We have compared the calculated values with data, and determined the compatibility between them by performing a χ^2 statistical analysis. Since the theoretical expressions depend on the mixing angle $\sin^2 2\phi$, the square mass difference δm^2_{14} (normal mass hierarchy) or δm^2_{34} (inverse mass hierarchy), and the baryonic density $\Omega_B h^2$, we have adopted the LSND limits on the mixing angle and the WMAP results on the baryonic density, as constraints.

The mixing between active neutrino mass eigenstates ν_i ($i = 1, 2, 3$), leading to neutrinos of a given flavor ν_k ($k =$ light, medium, heavy), is described by the mixing matrix U [25]

$$U = \begin{pmatrix} c_{13}c_{12} & s_{12}c_{13} & s_{13} \\ -s_{12}c_{23} - s_{23}s_{13}c_{12} & c_{23}c_{12} - s_{23}s_{13}s_{12} & s_{23}c_{13} \\ s_{23}s_{12} - s_{13}c_{23}c_{12} & -s_{23}c_{12} - s_{13}s_{12}c_{23} & c_{23}c_{13} \end{pmatrix}, \qquad (2)$$

where c_{ij} and s_{ij} stand for $\cos\theta_{ij}$ and $\sin\theta_{ij}$, respectively, and CP conservation is assumed. To this mixing we add the mixing of a sterile neutrino with: a) the neutrino mass eigenstate of lowest mass in the normal mass hierarchy, ν_1, and b) to the one of the inverse mass hierarchy, ν_3, by defining the mixing angle ϕ, such that the new mixing matrix U is redefined as $U(\phi)$ [17].

The mixing between neutrino mass eigenstates, and particularly the inclusion of the sterile neutrino as a partner of the light neutrino, affects the statistical occupation factors of neutrinos of a given flavor. The equation which determines the structure of the neutrino occupation factors, in the basis of mass eigenstates and for an expanding Universe, can be written:

$$\left(\frac{\partial f}{\partial t} - \mathrm{H}E_\nu \frac{\partial f}{\partial E_\nu}\right) = \imath\,[H_0, f], \tag{3}$$

where t is time, H is the expansion rate of the Universe, defined as $\mathrm{H} = \sqrt{\frac{4\pi^3 N}{45 M^2_{\text{Planck}}}}T^2 = \mu_P T^2$, T is the temperature, E_ν is the energy of the neutrino, and H_0 is the unperturbed mass term of the neutrino's Hamiltonian in the rest frame. The initial condition is fixed by defining the occupation numbers at the temperature $T_0 = 3$ MeV [26],

$$\left.\begin{pmatrix} f_{11} & f_{12} & f_{13} & f_{14} \\ f_{21} & f_{22} & f_{23} & f_{24} \\ f_{31} & f_{32} & f_{33} & f_{34} \\ f_{41} & f_{42} & f_{43} & f_{44} \end{pmatrix}\right|_{T_0} = \frac{1}{1+e^{E_\nu/T_0-\eta}} \begin{pmatrix} \cos^2\phi & 0 & 0 & \sin\phi\cos\phi \\ 0 & 1 & 0 & 0 \\ 0 & 0 & 1 & 0 \\ \sin\phi\cos\phi & 0 & 0 & \sin^2\phi \end{pmatrix}, \tag{4}$$

for the normal mass hierarchy, and

$$\left.\begin{pmatrix} f_{11} & f_{12} & f_{13} & f_{14} \\ f_{21} & f_{22} & f_{23} & f_{24} \\ f_{31} & f_{32} & f_{33} & f_{34} \\ f_{41} & f_{42} & f_{43} & f_{44} \end{pmatrix}\right|_{T_0} = \frac{1}{1+e^{E_\nu/T_0-\eta}} \begin{pmatrix} 1 & 0 & 0 & 0 \\ 0 & 1 & 0 & 0 \\ 0 & 0 & \cos^2\phi & \sin\phi\cos\phi \\ 0 & 0 & \sin\phi\cos\phi & \sin^2\phi \end{pmatrix}, \tag{5}$$

for the inverse mass hierarchy.

To obtain the solutions of Eq.(3) we have written the commutator in the r.h.s of Eq.(3), in terms of the square mass differences, $\delta m^2_{ij} = m^2_i - m^2_j$:

$$[H_0, f] = \frac{1}{2p}\begin{pmatrix} 0 & \delta m^2_{12} f_{12} & \delta m^2_{13} f_{13} & \delta m^2_{14} f_{14} \\ -\delta m^2_{12} f_{21} & 0 & \delta m^2_{23} f_{23} & \delta m^2_{24} f_{24} \\ -\delta m^2_{13} f_{31} & -\delta m^2_{23} f_{32} & 0 & \delta m^2_{34} f_{34} \\ -\delta m^2_{14} f_{41} & -\delta m^2_{24} f_{42} & -\delta m^2_{34} f_{43} & 0 \end{pmatrix}. \tag{6}$$

The value of the mixing angle θ_{13} is constrained by the upper limit given by [27] so that $\tan\theta_{13} \le 10^{-3}$. The solution in the basis of mass eigenstates reads

$$\begin{aligned} f_{ii} &= \frac{\text{const}}{1+e^{E_\nu/T-\eta}} \\ f_{ij} &= \frac{\text{const}}{1+e^{E_\nu/T-\eta}} \exp\left[\imath \frac{\delta m^2_{ij}}{6\mu_P}\frac{T}{E_\nu}\left(\frac{1}{T^3} - \frac{1}{T_0^3}\right)\right] \end{aligned} \tag{7}$$

where the normalization constants are fixed by the initial conditions ($T = T_0$). In the above expressions, η is the ratio between the neutrino chemical potential and the temperature. This parameter depends on the adopted value of the leptonic number L [14, 13]. Explicit expressions of η versus L can be found in [14]. In the present context we have taken η as an input for the calculations.

3.1. Decay Rates and neutron abundance

In the following we shall outline the main steps of the calculation of neutron decay rates, for the electroweak processes $n + e^+ \to p + \overline{\nu}$ and $n + \nu \to p + e^-$. The starting point is the calculation of the reduced rates $\lambda_\pm$

$$\lambda\left(n + \nu \to p + e^-\right) = \lambda_- = \lambda_0 \int_0^\infty dp_\nu p_\nu E_\nu p_e E_e \left(1 - f_e\right) f_l, \tag{8}$$

$$\lambda\left(n + e^+ \to p + \overline{\nu}\right) = \lambda_+ = \lambda_0 \int_0^\infty dp_e p_\nu E_\nu p_e E_e \left(1 - f_l\right) f_e, \tag{9}$$

and the total neutron to proton decay rate

$$\lambda_{np}(y) = \lambda_-(y) + \lambda_+(y). \tag{10}$$

The final expression for the neutron to proton decay rate is obtained by fixing the normalization λ_0, of Eq.(10), from the neutron half-life

$$\frac{1}{\tau} = \frac{4\lambda_0 \Delta m_{np}^5}{255}. \tag{11}$$

The neutron abundance, until the freeze-out of weak interactions, is expressed in terms of the neutron to proton decay rate, λ_{np} of Eq.(10) as

$$X_{\text{neutrons}} = \int_0^\infty dw\, e^{w+\eta} \left(\frac{1}{1 + e^{w+\eta}}\right)^2 e^{-\left(\mu_P \Delta m_{np}^2\right)^{-1} \int_w^\infty du u\left(1+e^{-u-\eta}\right)\lambda_{np}(u)}. \tag{12}$$

The quantity X_{neutrons} is, therefore, a function of λ_{np} and, consequently, of the occupation factors f_l, which contain the information about the mixing between active and sterile neutrinos. The next step consists on the calculation of primordial nuclear abundances. It is a semi-analytic approach based on the balance between production and destruction of a given nuclear element, which requires the knowledge of X_{neutrons}. The above presented framework shows that the calculation of primordial abundances may indeed be taken as a tool to test leptonic mechanisms, like the mixing between sterile and active neutrinos, as it has been pointed out by Kishimoto et al. [13].

To perform the calculations we have adopted the oscillation parameters determined from SNO, SK and CHOOZ measurements [27]. The mixing with the sterile neutrino, represented by the mixing angle ϕ, is taken as an unknown variable, within the limits fixed by the LSND data [23]. The mass splitting δm_{14}^2 (or δm_{34}^2) was taken from the analysis given by Keränen et al. [17]. The actual value is fixed at $\delta m^2 = 10^{-11}$ eV^2. We have then calculated the neutron abundance, by applying the formalism of the previous section. The baryonic density $\Omega_B h^2$ (see Ref.[29]) was varied within the limits $0.010 < \Omega_B h^2 < 0.035$. Concerning the value of η we have varied it in the interval determined by the allowed values of the potential lepton number, $\mathcal{L} = 2\mathrm{L}_{\nu_e} + \mathrm{L}_{\nu_\mu} + \mathrm{L}_{\nu_\tau}$, that is $0.0 \le \mathcal{L} \le 0.4$ [13, 14]. In the present calculations we have adopted the values $0.0 \le \eta \le 0.07$ which are consistent with the densities $0.0 \le \mathrm{L}_{\nu_e} \le 0.05$ [14].

To determine the allowed values of the mixing angle ϕ we have performed a χ^2-minimization, after computing the primordial abundances. The absolute minimum is located at $\sin^2 2\phi = 0.000 \pm 0.026$, and $\Omega_B h^2 = 0.0253 \pm 0.0015$, both set of results have been obtained by using the solution (7) for the occupations. The smallness of the mixing angle does not contradict LSND results but the value of the baryonic density is outside the limits determined by WMAP [24], that is: $(\Omega_B h^2)_{\text{WMAP}} = 0.0223 \pm 0.0008$. This disagreement between theory and data may be caused by large uncertainties in the ^{7}Li-data. As pointed out by Richard et al. [30], the validity of the data on ^{7}Li may be questioned by the uncertainties inherent to the physics of ^{7}Li in the interior of the stars, i.e; the turbulent transport in the radiative zone of stars. In contrast, the situation improves if the data on ^{7}Li are removed at the time of performing the statistical analysis. For this case, the best value of the mixing angle is $\sin^2 2\phi = 0.018 \pm 0.098$, and the baryonic density corresponding to the minimum, $\Omega_B h^2 = 0.0216 \pm 0.0017$, is indeed consistent with the WMAP data. The anomalous feature associated with the inclusion of ^{7}Li in the set of data persists if other elements are removed from the data. We have verified it by systematically removing, one at the time, the abundances of D, ^{3}He, and ^{4}He, and keeping the data on ^{7}Li.

For the case of inverse mass hierarchy the occupation factor is strongly constrained by the value of θ_{13} and the difference with respect to the thermal occupation factor vanishes. For both the normal and inverse hierarchy solutions, particle number conservation was enforced, on the average, by the factor μ_P (see its definition following Eq.(3)). Because of the high temperature we have not included collision terms in Eq. (3).

Similar results, related to the abundance of ^{7}Li, have been obtained in the calculations of nuclear abundances in the context of cosmological models [31], and also in the case of a two neutrino mixing [28].

Finally, the best values of the baryon density and the mixing angle, both with and without including ^{7}Li in the analysis, are shown in Table 4 as functions of η. In agreement with the expectations of [18], and with our owns, both sets of results do not differ much, or at least they do not show a pronounced dependence, with respect to the chemical potential.

Table 4. Best values of the mixing angle and of the baryonic density, determined from the χ^2 analysis of the calculated abundances, as functions of the parameter η. Left and right sides of the table show the results obtained with and without considering the data on ^{7}Li, respectively.

	All data		All data but ^{7}Li	
η	$\Omega_B h^2$	$\sin^2 2\phi$	$\Omega_B h^2$	$\sin^2 2\phi$
0.00	0.0253 ± 0.0015	0.000 ± 0.026	0.0216 ± 0.0017	0.018 ± 0.098
0.01	0.0250 ± 0.0014	0.000 ± 0.010	0.0216 ± 0.0020	0.002 ± 0.022
0.02	0.0248 ± 0.0014	0.000 ± 0.015	0.0218 ± 0.0020	0.004 ± 0.030
0.03	0.0246 ± 0.0012	0.000 ± 0.034	0.0216 ± 0.0018	0.008 ± 0.080
0.04	0.0244 ± 0.0016	0.000 ± 0.039	0.0216 ± 0.0019	0.018 ± 0.090
0.05	0.0244 ± 0.0016	0.000 ± 0.056	0.0216 ± 0.0017	0.052 ± 0.090
0.06	0.0244 ± 0.0016	0.000 ± 0.101	0.0216 ± 0.0018	0.108 ± 0.090

4. Conclusions

So far, the results presented in the previous sections show that it is indeed possible to combine nuclear, neutrino and cosmological models to analyze astrophysical data. The analysis of BBN results shows that there is a clear sensitivity of the data respect to neutrino mass hierarchies and to the mixing with sterile neutrinos. The calculated dependence of the baryon density $\Omega_B h^2$, respect to the mixing angle $\sin^2 2\phi$, is in good agreement with the limits extracted from the WMAP and LSND measurements, only if data on the abundance of ^{7}Li are excluded from the analysis. Changes in the vacuum expectation value of the Higgs field could explained for the discrepancies between the theoretical and observed BBN abundances.

Acknowledgments

This work was partly supported by the CONICET and by the ANPCYT (Argentina).

References

[1] Tzanavaris P *et al* 2007 *Mon. Not. Roy. Astrom. Soc.* **374** 634
[2] Malec A L *et al* 2010 *Mon. Not. Roy. Astrom. Soc.* **403** 1541
[3] Mosquera M E and Civitarese O 2010 *Astron. & Astrophys.* **520** A112
[4] Civitarese O, Moline M A and Mosquera M E 2010 *Nuc. Phys. A* **846** 157
[5] Flambaum V V and Wiringa R B 2007 *Phys. Rev. C* **76** 054002
[6] Berengut J C, Flambaum V V and Dmitriev V F 2010 *Phys. Lett. B* **683** 114
[7] Kawano L 1992 *FERMILAB-PUB-92-004-A*
[8] Ivanchik A *et al* 2010 *Mon. Not. Roy. Astrom. Soc.* **404** 1583
[9] Izotov Y I and Thuan T X 2010 *Astrophys. J.* **710** L67
[10] Hosford A *et al* 2009 *Astron. & Astrophys.* **493** 601
[11] Yao *et al* 2006 *J. Phys. G: Nucl. Part. Phys.* **33** 1
[12] Spergel D N *et al* 2007 *Astrophys. J. Suppl. Ser.* **170** 377
[13] Kishimoto C T, Fuller G M and Smith C J 2006 *Phys. Rev. Lett.* **97** 141301
[14] Smith C J, Fuller G M, Kishimoto C T and Abazajian K N 2006 *Phys. Rev. D* **74** 085008
[15] Cirelli M 2004 *Preprint* astro-ph/0410122
[16] Khlopov M Yu 1999 *Cosmoparticle Physics* (Singapore: World Scientific)
[17] Keränen P, Maalampi J, Myyryläinen M and Riitinen J 2003 *Phys. Lett. B* **574** 162
[18] Bernstein J, Brown L and Feinberg G 1989 *Rev. Mod. Phys.* **61** 25-39
[19] Abazajian K, Bell N F, Fuller G M and Wong Y Y Y 2005 *Phys. Rev. D* **72** 063004
[20] Bell N F, Volkas R R and Wong Y Y Y 1999 *Phys. Rev. D* **59** 113001
[21] Hannestad S 2003 *J. Cosmol. Astropart. Phys.* **05** 004
[22] Seljak U *et al* 2005 *Phys. Rev. D* **71** 103515
[23] McGregor G 2003 *Particle Physics and Cosmology (AIP Conf. Proc. vol 655)* ed Nieves J F and C N Leung (New York: AIP) 58
[24] Spergel D *et al* 2007 *Astrophys. J. Suppl.* **170** 377
[25] Bandyopadhyay A, Choubey S, Goswami S and Kar K 2002 *Phys. Rev. D* **65** 073031
[26] Dolgov A D, Hansen S H and Semikoz D V 1997 *Nucl. Phys. B* **503** 426-444
[27] Ahmad Q R *et al* 2001 *Phys. Rev. Lett.* **87** 071301
Fukuda S *et al* 2001 *Phys. Rev. Lett.* **86** 5651
Appollonio M *et al* 1999 *Phys. Lett. B* **466** 415
[28] Civitarese O and Mosquera M E 2008 *Int. J. Mod. Phys. E* **17** 351
[29] Civitarese O and Mosquera M E 2008 *Phys. Rev. C* **77** 045806
[30] Richard O, Michaud G and Richer J 2005 *Astrophys. J.* **619** 538
[31] Cuoco A *et al* 2004 *Int. J. Mod. Phys. A* **19** 4431

XXXIV Symposium on Nuclear Physics IOP Publishing
Journal of Physics: Conference Series **322** (2011) 012006 doi:10.1088/1742-6596/322/1/012006

Nuclear reactions for astrophysics and other applications

J E Escher, J T Burke, F S Dietrich, N D Scielzo and J J Ressler
Lawrence Livermore National Laboratory, Livermore, CA 94550, USA
E-mail: escher1@llnl.gov

Abstract. Cross sections for compound-nuclear reactions are required for many applications. The surrogate nuclear reactions method provides an indirect approach for determining cross sections for reactions on unstable isotopes, which are difficult or impossible to measure otherwise. Current implementations of the method provide useful cross sections for (n,f) reactions, but need to be improved upon for applications to capture reactions.

1. Introduction

Compound nuclear reactions play an important role in many applications. Their cross sections are required input for astrophysical models that describe stellar evolution and nucleosynthesis and for modeling processes that are relevant to generating energy. Cross sections for neutron capture, neutron-induced fission, and other processes are required. Often the reactions of interest involve short-lived or highly radioactive target nuclei, which makes it difficult or impossible to measure their cross sections directly. To determine the cross sections of interest, indirect methods, such as inverse-kinematics experiments with radioactive beams, combined with theoretical treatments, are becoming increasingly important. This paper discusses recent progress of the surrogate nuclear reaction method, an indirect approach for determining cross sections for nuclear reactions $a + A \rightarrow B^* \rightarrow c + C$, that proceed through the formation of a compound nucleus (B^*). The method can be applied when the target is too short-lived to allow for a direct measurement, but a projectile-target combination can be found so that the compound nucleus of interest, B^*, can be produced in an alternative reaction $d + D \rightarrow B^* + b$.

In Sec. 2, representative examples from the fields of nuclear astrophysics and nuclear energy are given to illustrate the importance of cross sections for reactions on unstable targets. In Section 3 an outline of the surrogate approach is given. Some examples of successful applications of the method to (n,f) reactions are shown in Sec. 4. The limitations of presently-employed approximations are illustrated in Sec. 5 for capture reactions.

2. Compound reactions for astrophysics and nuclear energy

Cross sections for compound-nuclear reactions are required input for astrophysical models and nuclear-energy simulations.

2.1. Nucleosynthesis of heavy elements

An important goal of nuclear astrophysics research is to explain the origin of the heavy elements (A>56). We have acquired a basic, but incomplete, understanding of the processes that generate the energy in stars such as our sun, that drive stellar evolution and that are responsible for the synthesis of the elements [1]. Nucleosynthesis of heavy elements beyond ^{56}Fe is known to take place primarily by neutron capture on lighter seed nuclei in the s (slow neutron-capture) and r (rapid neutron-capture) processes [1, 2, 3], with other processes contributing to the abundances of some specific isotopes. Models of these processes predict relative isotopic abundances for specific astrophysical scenarios. They require nuclear physics input and are constrained by measured isotopic abundances.

The s process proceeds via neutron captures and beta decays through nuclides in and very near the valley of stability. Of particular interest to models of this process are neutron captures on s-process branch points, unstable nuclei with a life time long enough to allow the s process to proceed by either neutron capture or β decay. The strength with which one path dominates over the other depends on environmental variables, such as neutron density, temperature, and pressure, as well as on nuclear properties, specifically capture rates and beta-decay life times. Information on the astrophysical conditions can be inferred if the nuclear properties are known.

Beyond being important for gaining a more detailed description of the s process environments, reliable s-process abundances are crucial for our understanding of the r (rapid neutron-capture) process. The r process takes place in an environment with high temperature and high neutron flux. In such conditions the average time between neutron captures is much shorter than the life time for β-decay and reaction flows can proceed to very neutron-rich nuclei. Open questions include the exact path along the nuclear chart of the neutron captures and beta decays involved and the astrophysical site(s) of the process. Typically, r-process abundances are inferred by subtracting calculated s-process abundances from measured total abundances. This requires detailed calculations to predict s-process abundances, and reliable cross sections for neutron captures along the s-process path. Beta-decay half lives and nuclear masses, which determine nucleon separation energies, are considered to be the most important nuclear physics inputs for r-process models, as they determine the neutron capture path in scenarios that assume (n,γ) and (γ,n) reactions to be in equilibrium with each other, but models that do not invoke the equilibrium assumption require neutron capture rates for thousands of unstable nuclei.

2.2. Nuclear energy

The world-wide growing demand for energy, coupled with increasing concern over pollution resulting from burning of fossil fuels, has resulted in renewed interest in nuclear energy. To exploit nuclear energy in a clean, efficient, and safe manner, concerns regarding reactor safety, waste handling, proliferation risks, and economic competitiveness have to be addressed.

Advanced nuclear-energy systems may recycle actinides produced in conventional reactors or take advantage of alternative fuel cycles. A thorium-based fuel cycle, for instance, is appealing, since it produces less radiotoxic heavy transurianum isotopes than the conventional uranium-plutonium fuel cycle; thorium is also more abundant in the earth's crust than uranium. An important component of the research and development for these reactor concepts is the improvement of the fundamental nuclear data [4]. The transuranic nuclides, for example, play a much more prominent role in these new designs and yet the available cross-section data is quite limited. In addition, neutron-capture reactions could be used to incinerate long-lived fission fragments and therefore limit the challenges associated with the reactor waste.

Sensitivity studies [5] indicate that high-quality, reliable cross-section data are needed for neutron-induced reactions for a wide variety of radioactive isotopes covering neutron energies from thermal up to tens of MeV. In particular, capture and fission reactions on many of the isotopes of thorium, uranium, plutonium, and the minor actinides (such as ^{237}Np, $^{241-243}$Am, and $^{244-245}$Cm), as well as certain long-lived fission fragments. Many of these cross sections are extremely challenging to measure directly but significantly contribute to uncertainties in the reliable design and safe operation of a nuclear-energy system.

Some relevant cross sections, such as the (n, f) cross sections for ^{241}Am, $^{243-244}$Cm have already been determined with the surrogate approach [6, 7, 8, 9, 10]. It is anticipated that the surrogate reaction method will continue to provide valuable cross-section results on isotopes for which there is limited, poor-quality, or no data of use for nuclear-energy applications.

3. The surrogate idea

The appropriate formalism for the description of a compound-nuclear reaction $a + A \rightarrow B^* \rightarrow c + C$ is a statistical one [11]. Based on Bohr's hypothesis of the independence of formation and decay of the compound nucleus (CN), the cross sections are calculated in the framework of the Hauser-Feshbach formalism, which properly takes account of the conservation of angular momentum and parity in the reaction:

$$\sigma_{\alpha\chi}(E_a) \;=\; \sum_{J,\pi} \sigma_\alpha^{CN}(E_{ex}, J, \pi)\; G_\chi^{CN}(E_{ex}, J, \pi)\,, \tag{1}$$

with α and χ denoting the relevant entrance and exit channels, $a + A$ and $c + C$, respectively. The excitation energy E_{ex} of the compound nucleus, B^*, is related to the center-of-mass energy E_a in the entrance channel via the energy needed for separating a from B: $E_a = E - S_a(B)$. In many cases the formation cross section $\sigma_\alpha^{CN} = \sigma(a + A \rightarrow B^*)$ can be calculated to a reasonable accuracy by using optical potentials, while the theoretical decay probabilities G_χ^{CN} for the different decay channels χ are often quite uncertain. The latter are difficult to calculate accurately since they require knowledge of optical models, level densities, and strength functions for the various possible exit channels. The objective of the surrogate method is to determine or constrain these decay probabilities experimentally.

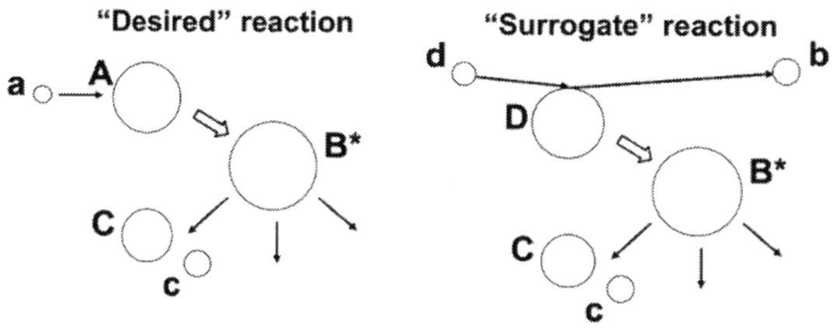

Figure 1. Schematic representation of the "desired" (left) and "surrogate" (right) reaction mechanisms. The basic idea of the surrogate approach is to replace the first step of the desired reaction, $a + A$, by an alternative (surrogate) reaction, $d + D \rightarrow b + B^*$, that populates the same compound nucleus. The subsequent decay of the compound nucleus into the relevant channel, $c + C$, can then be measured and used to extract the desired cross section.

In the surrogate approach, the compound nucleus B^* is produced by means of an alternative ("surrogate"), direct reaction, $d + D \rightarrow b + B^*$, and the desired decay channel $\chi(B^* \rightarrow c + C)$ is observed in coincidence with the outgoing particle b (see Fig. 1). The probability for forming B^* in the surrogate reaction (with specific values for E_{ex}, J, π) is $F_\delta^{CN}(E_{ex}, J, \pi)$, where δ refers to the entrance channel reaction $D(d, b)$. The quantity

$$P_{\delta\chi}(E_{ex}) = \sum_{J,\pi} F_\delta^{CN}(E_{ex}, J, \pi)\; G_\chi^{CN}(E_{ex}, J, \pi)\,, \tag{2}$$

which gives the probability that the compound nucleus B^* was formed with energy E_{ex} and decayed into channel χ, can be obtained experimentally, by measuring N_δ, the total number of surrogate events, and $N_{\delta\chi}$, the number of coincidences between the direct-reaction particle and the observable that identifies the relevant exit channel: $P^{exp}_{\delta\chi}(E_{ex}) = N_{\delta\chi}/N_\delta \epsilon_\delta$. Here, ϵ_δ denotes the efficiency for detecting the outgoing direct-reaction particle b and the exit channel χ. To simplify the notation, we suppress the dependence of the coincidence probability $P_{\delta\chi}(E_{ex})$ on the angle θ_b of the outgoing direct-reaction particle b.

The distribution $F^{CN}_\delta(E_{ex}, J, \pi)$, which may be very different from the CN spin-parity populations following the absorption of the projectile a in the desired reaction, has to be determined theoretically, so that the branching ratios $G^{CN}_\chi(E_{ex}, J, \pi)$ can be extracted from the measurements. In practice, the decay of the CN is modeled and the $G^{CN}_\chi(E_{ex}, J, \pi)$ are obtained by adjusting parameters in the model to reproduce the measured probabilities $P_{\delta\chi}(E_{ex})$ [12, 13]. Subsequently, the sought-after cross section can be obtained by combining the calculated cross section $\sigma^{CN}_\alpha(E_{ex}, J, \pi)$ for the formation of B^* (from $a+A$) with the extracted decay probabilities $G^{CN}_\chi(E_{ex}, J, \pi)$ for this state, see Eq. 1.

Weisskopf-Ewing Approximation. Under certain circumstances, the decay of the intermediate equilibrated system, the compound nucleus, becomes independent of its angular momentum and parity, and the cross section for the reaction factorizes into a simple product of a formation cross section and a decay probability for the exit channel of interest [11]:

$$\sigma_{\alpha\chi}(E_a) \quad = \quad \sigma^{CN}_\alpha(E_{ex}) \ \mathcal{G}^{CN}_\chi(E_{ex},) \ , \qquad (3)$$

where $\sigma^{CN}_\alpha(E_{ex}) = \sum_{J,\pi} \sigma^{CN}_\alpha(E_{ex}, J, \pi)$ is the reaction cross section describing the formation of the compound nucleus in the desired reaction and $\mathcal{G}^{CN}_\chi(E_{ex})$ denotes the J^π-independent decay probability for the exit channel χ. This is the Weisskopf-Ewing (WE) limit of the Hauser-Feshbach theory [11]. In the context of surrogate reactions, the Weisskopf-Ewing approximation greatly simplifies the application of the method: It becomes straightforward to obtain the J^π-independent branching ratios $\mathcal{G}^{CN}_\chi(E_{ex})$ from measurements of $P_{\delta\chi}(E_{ex})$ [$= \mathcal{G}^{CN}_\chi(E_{ex})$ since $\sum_{J,\pi} F^{CN}_\delta(E_{ex}, J, \pi) = 1$] and to calculate the desired reaction cross section. Calculating the direct-reaction probabilities $F^{CN}_\delta(E_{ex}, J, \pi)$ and modeling the decay of the compound nucleus are no longer required. Most applications to date invoke the Weisskopf-Ewing approximation.

4. Neutron-induced fission reactions

The surrogate approach was first employed in the 1970s to estimate neutron-induced fission cross sections from transfer reactions. These early applications of the method made use of the Weisskopf-Ewing approximation: The transfer reactions of the 1970s [14, 15] produced (n,f) cross section estimates for various actinide targets which agreed with direct measurements (where available) to about 10-20% for incident neutron energies above 1 MeV. Discrepancies at lower energies were later attributed to large uncertainties in the low-energy optical model employed, and the use of the Weisskopf-Ewing approximation [12, 13].

More recently, a French group has carried out surrogate experiments at the Institut de Physique Nucléaire (IPN) in Orsay to determine cross sections for neutron-induced reactions on several minor actinide nuclei relevant to the thorium-uranium fuel cycle and the transmutation of nuclear waste [6, 7, 10]. The transfer reactions ^{232}Th(^{3}He,x) and ^{243}Am(^{3}He,x), with x=α,t,d,p, were employed to obtain (n,f) and (n,γ) cross sections for Th and Pa targets [6, 7] and (n,f) cross sections for Cm and Am targets [10], respectively. The analyses assumed that the Weisskopf-Ewing approximation is valid. The extracted (n,f) cross sections were found to be consistent with known directly-measured cross section measurements, where these were available. They proved useful for resolving controversies between discrepant measurements and for providing data for previously unavailable energy regions.

The STARS/LiBerACE collaboration in the United States has carried out a number of experiments at the 88-inch cyclotron at Lawrence Berkeley National Laboratory. Light-ion beams have been used for inelastic scattering, charge exchange, and one- or two-neutron transfer reactions. Fission cross sections, e.g. the ^{237}Np(n,f) cross section [16], have been determined using the Weisskopf-Ewing approximation or a variant thereof, the surrogate ratio method.

The *Surrogate Ratio approach* [17, 18, 19] requires the (approximate) validity of the Weisskopf-Ewing limit. The ratio $R(E) = \sigma_{\alpha_1\chi_1}/\sigma_{\alpha_2\chi_2}$ of the cross sections of two CN reactions is measured, using two surrogate experiments. An independent determination of the cross section $\sigma_{\alpha_2\chi_2}$ can then be used to deduce $\sigma_{\alpha_1\chi_1}$. An advantage of using the ratio method is the fact that it eliminates the need to accurately measure the total number of surrogate reaction events (N_δ), since one determines the ratio of coincidence probabilities $P_{\delta_1\chi_1}/P_{\delta_2\chi_2}$, rather than an absolute probability $P_{\delta\chi} = N_{\delta\chi}/N_\delta$. Furthermore, there are indications that small to moderate deviations from the WE limit cancel in this approach [17]. Cross sections for (n,f) reactions extracted in the ratio approximation have been tested for consistency with results from direct measurements, complementary surrogate experiments [19, 20], and theoretical simulations [17]. For (n,f) reactions one typically finds that the spin-parity mismatch between the desired and surrogate reactions has a much smaller effect on the extracted cross section than in an approach that uses the WE approximation and absolute probabilities. Also, deviations caused by pre-equilibrium effects are diminished, thus improving the overall agreement between extracted and expected cross sections.

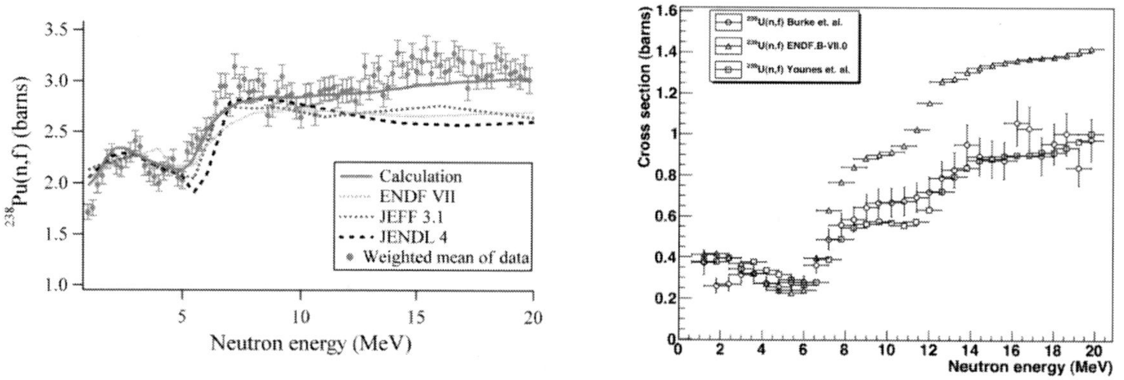

Figure 2. Fission cross sections obtained from surrogate ratio measurements. The ^{238}Pu(n,f) cross section (left panel) was determined relative to the known ^{234}U(n,f) and ^{235}U(n,f) cross sections [24]. The weighted average of the two measurements is compared to several evaluations; the solid line represents a calculation based on the new data. The ^{239}U(n,f) cross section (right panel) was determined relative to the known ^{235}U(n,f) cross section [21]; here it is compared to the result of an earlier surrogate measurement [22] and to the ENDF/B-VII evaluation [23].

Two recent applications of the ratio approach are shown in Figure 2. The ^{238}Pu(n,f) cross section, which is needed for reactor applications and for transmutation studies, was recently determined using inelastic scattering surrogate reactions [24]. Ressler *et al* [24] produced the compound nuclei ^{239}Pu*, ^{235}U*, and ^{236}U* via inelastic α scattering. Surrogate ratio analyses yielded the desired ^{238}Pu(n,f) cross section relative to both the ^{234}U(n,f) and the ^{235}U(n,f) cross sections. Since the latter two are known, the ^{238}Pu(n,f) cross section could be extracted; the weighted average of both measurements is shown in the left panel of Figure 2. The two measurements are in good agreement with each other for 5-20 MeV and the averaged cross section agrees well with previous, direct, measurements in the 5-10 MeV range; it is somewhat higher (by less than 20%) than those at about 15 MeV. The ratio measurement does not, and is not expected to, produce highly-accurate fission results at low energies (< 5 MeV here), due to the underlying Weisskopf-Ewing assumption. It does, however, provide continuous data for neutron energies from 5 to 20 MeV, and supplement earlier measurements, which were sparse in the energy regime around 10-15 MeV.

The right panel of Figure 2 shows the ^{239}U(n,f) cross section (open circles), which was recently determined relative to the known ^{235}U(n,f) cross section [21]. The CN ^{240}U* and ^{236}U* were produced via the two-neutron transfer reactions ^{238}U(^{18}O,^{16}O) and ^{234}U(^{18}O,^{16}O), respectively, and ^{16}O – fission fragment coincidence events were counted. The extracted ^{239}U(n,f) cross section agrees well with the result of an earlier surrogate experiment (open squares, from Ref. [22]); both are lower than the ENDF/B-VII evaluation (open triangles). No direct measurements exist, since the half life of ^{239}U is only 23.4 minutes.

The literature and the above examples indicate that (n,f) cross sections extracted from surrogate data are typically consistent with direct measurements (where available) and/or other surrogate measurements, despite the approximations used. In addition, calculations, which test the approximation schemes employed in the analyses of surrogate fission data [17], illustrate the level of accuracy that one can under reasonable circumstances expect from the surrogate approach. Discrepancies between indirectly and directly measured cross sections are often less than 10%. The largest deviations are found at low energies, where the Weisskopf-Ewing approximation is not expected to be valid; in those cases, it becomes necessary to account for the differences in the spin-parity distributions ocurring in the desired and surrogate reactions.

5. Capture reactions

Capture cross sections provide specific challenges for the surrogate approach. First, the level of precision required for the cross section is often higher than in the fission case: Recent advances in modeling the astrophysical s process have resulted in requests to determine capture cross sections within a few percent and nuclear-energy applications require cross sections to within 5-10% [25, 5]. Achieving an accuracy of a few percent is challenging, but constraining an unknown (n,γ) cross section to within 20-30% should be considered a meaningful improvement of the situation, in particular since current cross section evaluations often show large deviations from each other. Secondly, it is the low-energy regime that is relevant to many applications. For s-process applications, for example, one needs cross sections from a few keV to about 200 keV. Both calculations and measurements have shown that this is the energy range for which the Weisskopf-Ewing approximation typically breaks down.

Theoretical studies have been carried out to assess the feasibility of obtaining capture cross sections from surrogate measurements and to determine promising candidates for such measurements. The strategy followed in these investigations is to extract information from Hauser-Feshbach calculations that have been adjusted to reproduce known cross sections (capture and, where applicable, fission). The branching ratios (channel probabilities), $G_\chi^{CN}(E, J, \pi)$ that enter Eqs. 1 and 2 can be calculated via this procedure; the Weisskopf-Ewing limit is reached when the branching ratios are approximately equal.

Overall, the studies show that the probability for a compound nucleus to decay via γ emission ($\chi = \gamma$) depends sensitively on the spin-parity population of the nucleus prior to decay. The dependence of the γ-branching ratios on the $J\pi$ distribution is greater than that found previously for fission ($\chi = fission$). Calculations for representative Zirconium, Gadolinium, and Uranium nuclei showed a strong dependence of the γ branching ratios on the spins populated in the compound nucleus. The effect was particularly strong for the ^{92}Zr nucleus, which has a closed proton subshell (Z=40) and a nearly-closed neutron shell (N=52 $\approx$ 50) [26]. A comparison with the results for Gadolinium and Uranium confirms the notion that the higher level densities present in the deformed rare-earth and actinide regions do reduce the sensitivity of the γ-decay probabilities to compound-nuclear spin-parity distributions and nuclear-structure effects [27].

For Gadolinium, we demonstrate in Fig. 3a that the (n,γ) cross sections obtained from a Weisskopf-Ewing analysis of surrogate data can differ significantly from the expected 'true' cross section. Shown are the results of a Weisskopf-Ewing analysis of the ^{156}Gd$(p, p'\gamma)$ coincidence data measured by Scielzo *et al* [28] (symbols with error bars) and theory results of a sensitivity study by Escher and Dietrich [27] (dashed and dotted curves). The ^{155}Gd(n,γ) cross section extracted from the indirect measurement is a factor of 2-3 larger than the directly-measured cross section (solid curve). It falls, for the most part, between the theoretical curves. The latter were obtained by combining calculated branching ratios and schematic spin-parity disctributions to simulate surrogate measurements, $P_{\delta\gamma}^{sim}(E) = \sum_{J,\pi} F_{\delta}^{CN}(E, J, \pi)\, G_{\gamma}^{CN}(E, J, \pi)$, and carrying out a WE 'analysis' to obtain the desired cross section, $\sigma_{n,\gamma}^{WE,sim}(E) = \sigma_{n}^{CN}(E) P_{\delta\gamma}^{sim}(E)$, where $\sigma_{n}^{CN}(E)$ denotes the CN formation cross section. The range of cross sections, $\sigma_{n,\gamma}^{WE,sim}(E)$, obtained by varying the simulated spin distributions within reasonable limits provides a measure

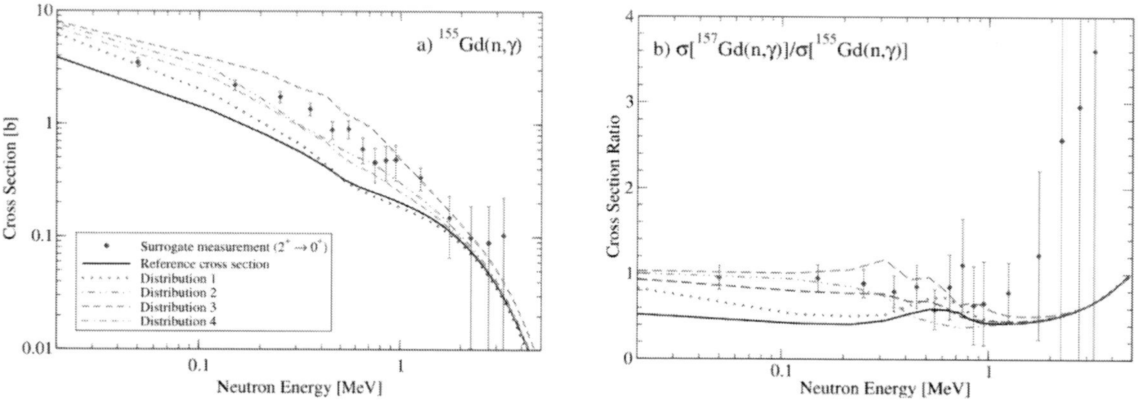

Figure 3. a) Weisskopf-Ewing estimate for the ^{155}Gd(n,γ) cross section. The cross section extracted from a surrogate ^{156}Gd$(p, p'\gamma)$ measurement [28] is compared to WE 'analyses' of simulated surrogate experiments [27], for different CN spin-parity distributions, and to the reference cross section, which is a fit to directly-measured cross sections. b) Cross section ratio obtained from (measured and simulated) surrogate data, compared to the ratio of evaluated cross sections. The spin-parity distributions used in the simulations are from Ref. [27], Fig. 10.

The studies of the Gadolinium region also show that the WE approximation overestimates the (n,γ) cross section by factors which depend on the nucleus under consideration [27, 28]. For the ^{155}Gd(n,γ) and ^{157}Gd(n,γ) cross sections, the differences are large, despite the structural similarities of the relevant nuclei. Consequently, the ratio approach results in ^{155}Gd(n,γ) cross sections that are too large by up to a factor two for energies below about $E_n = 0.7$ MeV (Fig. 3b). The effect is seen in both the theoretical sensitivity study and the experimental results. Overall, the ratio approach was found to reduce, but not eliminate, the effect of the spin-parity mismatch on the extracted cross sections for energies where the WE assumption is a poor approximation.

6. Concluding remarks

Indirect approaches have to be developed in order to provide much-needed nuclear data, in particular cross sections for reactions on unstable isotopes. We have reviewed recent application of the surrogate nuclear reaction method, which aims at providing cross section information for compound nuclear reactions. Past applications of the surrogate method have demonstrated that it can provide useful cross section estimates for neutron-induced fission of actinides. Similar success in applications to neutron capture for a range of isotopes would be very valuable and remains to be demonstrated. Most analyses of the fission data carried out so far have made approximations that are likely to break down in situations relevant for extracting (n,γ) cross sections from surrogate measurements, making this a more challenging reaction to tackle. The examples discussed show this breakdown for capture reactions on Gadolinium isotopes. They underscore the need to take into account the different spin-parity distributions that occur in the desired and surrogate reactions. Work to address this issue is underway [29]

Acknowledgments
This work was performed under the auspices of the U.S. Department of Energy by Lawrence Livermore National Laboratory under contract DE-AC52-07NA27344.

References

[1] Rolfs C E and Rodney S 1988 *Cauldrons in the Cosmos* (Chicago: University of Chicago Press)
[2] Burbidge E M, Burbidge G R, Fowler W A and Hoyle F 1957 *Rev. Mod. Phys.* **29** 547–650
[3] Wallerstein G, Iben I, Parker P *et al* 1997 *Rev. Mod. Phys.* **69** 995
[4] Colonna N, Belloni F, Berthoumieux E *et al* 2010 *Energy and Environmental Science* **3** 1910
[5] Aliberti G, Palmiotti G, Salvatores M *et al* J 2006 *Annals of Nuclear Energy* **33** 700
[6] Petit M, Aiche M, Barreau G *et al* 2004 *Nucl. Phys. A* **735** 347
[7] Boyer S, Dassie D, Wilson J *et al* 2006 *Nucl. Phys. A* **775** 175
[8] Nayak B K, Saxena A, Biswas D C *et al* 2008 *Phys. Rev. C* **78** 061602
[9] Goldblum B L, Stroberg S R, Allmond J M *et al* 2009 *Phys. Rev. C* **80** 044610
[10] Kessedjian G, Jurado B, Aiche M *et al* 2010 *Physics Letters B* **692** 297
[11] Gadioli E and Hodgson P E 1992 *Pre-Equilibrium Nuclear Reactions* (Oxford: Clarendon Press)
[12] Younes W and Britt H C 2003 *Phys. Rev. C* **67** 024610
[13] Younes W and Britt H C 2003 *Phys. Rev. C* **68** 034610
[14] Cramer J D and Britt H C 1970 *Nucl. Sci. and Eng.* **41** 177
[15] Britt H C and Wilhelmy J B 1979 *Nucl. Sci. and Eng.* **72** 222
[16] Basunia M, Clark R, Goldblum B *et al* 2009 *Nucl. Instr. and Meth. B* **267** 1899
[17] Escher J E and Dietrich F S 2006 *Phys. Rev. C* **74** 054601
[18] Plettner C *et al* 2005 *Phys. Rev. C* **71** 051602(R)
[19] Burke J T, Bernstein L A, Escher J *et al* 2006 *Phys. Rev. C* **73** 054604
[20] Lesher S R, Burke J T, Bernstein L A *et al* 2009 *Phys. Rev. C* **79** 044609
[21] Burke J T *et al* 2011 Surrogate two-neutron transfer reactions to populate ^{240}U to be submitted
[22] Younes W, Britt H C, Becker J A and Wilhelmy J B 2003 Initial estimate for the ^{237}U(n,f) cross section for $0.1 \leq E_n$ (MeV) ≤ 20 Tech. Rep. UCRL-ID-154194 Lawrence Livermore National Laboratory
[23] Young P G, Chadwick M B, MacFarlane R E *et al* 2007 *Nuclear Data Sheets* **108** 2589
[24] Ressler J J, Burke J T, Escher J E *et al* 2011 *Phys. Rev. C* **in press**
[25] Käppeler F and Mengoni A 2006 *Nuclear Physics A* **777** 291
[26] Forssén C, Dietrich F, Escher J, Hoffman R and Kelley K 2007 *Phys. Rev. C* **75** 055807
[27] Escher J E and Dietrich F S 2010 *Phys. Rev. C* **81** 024612
[28] Scielzo N D, Escher J E, Allmond J M *et al* 2010 *Phys. Rev. C* **81** 034608
[29] Escher J E, Burke J T, Dietrich F S, Scielzo N D, Thompson I J and Younes W 2011 *Rev. Mod. Phys.* **submitted**

Evaporation protons from $^{8}B+^{58}Ni$ at near barrier energies

P Amador-Valenzuela[1,2], E F Aguilera[1,*], E Martinez-Quiroz[1], D Lizcano[1], J J Kolata[3], A Roberts[3], F D Becchetti[4], M Ojaruega[4], M Febbraro[4], V Guimarães[5], E S Rossi Jr[6], J F P Huiza[7], L Acosta[8,9] and T L Belyaeva[2]

[1] Departamento de Aceleradores, Instituto Nacional de Investigaciones Nucleares, Apartado Postal 18-1027, C. P. 11801, México, D.F., México
[2] Universidad Autónoma del Estado de México, Código Postal 50000, Toluca, México
[3] Physics Department, University of Notre Dame, Notre Dame, In, 46556-5670, USA
[4] Physics Department , University of Michigan, Ann Arbor, Mi, 48109-1120, USA
[5] Instituto de Fisica, Universidade de Sao Paulo, P.O.Box 66318, 05389-970 Sao Paulo, SP, Brazil
[6] UNIFIEO - Centro Universitário FIEO - CEP 06020-190, Osasco-SP, Brazil
[7] Departamento de Química e Exatas, Universidade Estadual do Sudoeste da Bahia, 45206-190 Bahia, Brazil
[8] Departamento de Física Aplicada, Universidad de Huelva, E-21071 Huelva, Spain
[9] INFN, Laboratori Nazionali del Sud, I-95123 Catania, Italy

E-mail: *eli.aguilera@inin.gob.mx

Abstract. Yields of evaporated protons from the $^{8}B+^{58}Ni$ reaction are measured at backward angles, for several near barrier energies. Statistical model calculations using the code PACE are used to extrapolate the measurements to the whole angular region in order to get angle integrated cross sections. Fusion cross sections are deduced by using the calculated proton multiplicities. The obtained fusion excitation function shows a large enhancement as compared to BPM calculations using conventional barrier parameters.

1. Introduction

The short-lived radioactive nucleus ^{8}B is adjacent to the proton drip line and has a very small proton separation energy of only 0.138 MeV. It has attracted much attention because it may have a proton halo. Evidence for this, both theoretical and experimental, has appeared in the literature in recent years [1–10].

The present work is part of an effort to clarify the proton-halo wave function of ^{8}B as it relates to reaction cross sections at near barrier energies. For this purpose, studies of fusion, breakup and total yields are important. The kinematic regime near the barrier emphasizes peripheral reactions, where halo signatures are most probable to appear. For the $^{8}B+^{58}Ni$ system, the breakup measurement of Refs. [1, 2] did certainly show such a possible signature: Coulomb-nuclear interference at very large distances was shown to play an important role.

As for total reaction yields, very recently it was shown that the near- and sub-barrier reaction yields for ^{8}B show striking similarities with previous observations for the neutron-halo projectile

^{6}He. As an immediate antecedent to the present work, these results will be summarized in the next section.

2. Antecedents

For ^{6}He projectiles, large enhancements were observed below the barrier for a ^{209}Bi target [11–13], and also for targets closer to ^{58}Ni [14, 15]. For systems with lighter targets such as ^{6}He+^{12}C [16] and ^{6}He+^{27}Al [17], the neutron halo effects seemed to be smaller. We will first show the data recently obtained for the proton-halo projectile ^{8}B, and then a comparison with data corresponding to the neutron-halo projectile ^{6}He will be done.

2.1. Total reaction yields for ^{8}B projectiles

Elastic scattering angular distributions were reported for ^{8}B+^{58}Ni at five near barrier energies [18]. Optical model fits to the data were used to extract the corresponding total reaction cross sections. Since nuclei of ^{6}Li and ^{7}Be were also present in the beam in substantial amounts, data for these two projectiles were also obtained and analyzed. The ^{7}Be projectile is particularly interesting because it is the core of ^{8}B, so a comparison between these two projectiles should in principle give information about the reactions of the halo.

In order to make the comparison, reduced cross sections and reduced energies were used, according to the expressions $\sigma_{Reduced} = \sigma_R/(A_p^{1/3} + A_t^{1/3})^2$, $E_{Reduced} = E_{c.m.}/[Z_pZ_t/(A_p^{1/3} + A_t^{1/3})]$. This reduction of data is expected to eliminate trivial effects of size and charge, thus making data for different systems directly comparable.

Fig. 1 shows the comparison for the three projectiles. The halo system seems to define a single trajectory in this plot, while the data for ^{7}Be and ^{6}Li seem to group also in one single trajectory, different from the previous one. An important point is that in this reduced plot the trajectory for the halo nucleus lies above that for the weakly bound projectiles, which could be an actual signature for the presence of the halo. A comparison with additional systems reported in the literature seems to corroborate this conclusion, as described in next section.

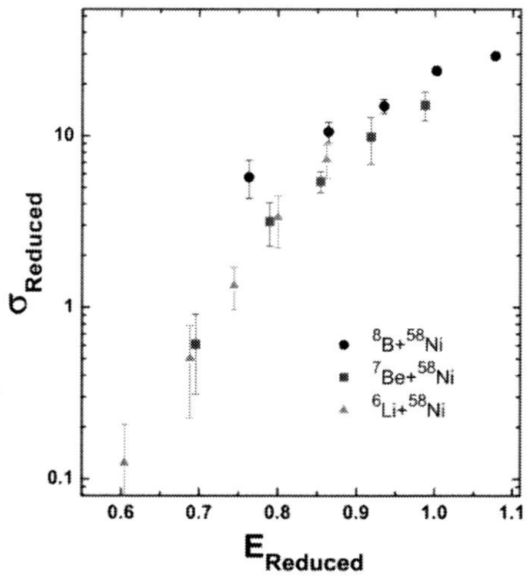

Figure 1. Total reaction cross sections for the (^{8}B,^{7}Be,^{6}Li)+^{58}Ni systems, plotted in reduced form. Data from Ref. [18].

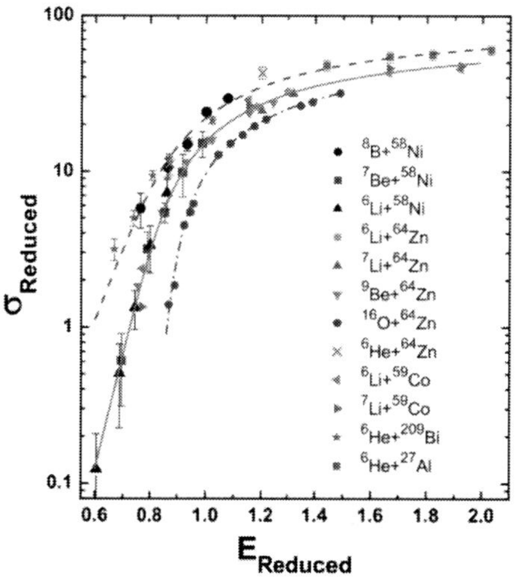

Figure 2. Comparison of reduced cross sections for twelve different systems. Figure adapted from Ref. [19].

Journal of Physics: Conference Series **322** (2011) 012007 doi:10.1088/1742-6596/322/1/012007

2.2. Comparison to other systems

In Ref. [19], the above results were compared to the reported data for other systems, which include the neutron-halo projectile ^{6}He as well as several isotopes of Li and Be, which are weakly bound. In addition, data for the strongly bound projectile ^{16}O were also included in the comparison.

We see in Fig. 2 that the reduced cross sections for all studied weakly bound (but otherwise normal) systems lie on the same trajectory when plotted as a function of reduced energy. A similar conclusion applies to the halo systems, but the respective trajectory lies above the previous one. The trajectory corresponding to the ^{16}O projectile, on the other hand, is the one lying lowest in the plot. In other words, if we characterize the trajectories by a mean value of the respective projectile binding energies, they lie progressively lower in the plot with increasing binding energy.

It was also shown [19] that the trajectories could be characterized by respective Wong-model [20] fits. These fits are illustrated with the curves in Fig. 2. The results of this figure seem to rule out previous conclusions [17] that for ^{6}He+^{27}Al, the neutron halo effects seem to be smaller. The most striking result is the fact that the reduced cross sections for the proton-halo nucleus ^{8}B and the two-neutron-halo nucleus ^{6}He follow identical trajectories despite differences in the structure, binding energy, and reaction mechanisms [18] of these systems. It is expected that the fusion measurements for ^{8}B will help further understand the peculiarities of the proton-halo system.

3. Experimental details

The ^{8}B beam was obtained from primary ^{6}Li projectiles through the ^{3}He(^{6}Li,n)^{8}B direct transfer reaction using the TwinSol radioactive nuclear beam facility at the University of Notre Dame [21]. The basic setup of the TwinSol facility consists of a production target chamber, where the primary reaction takes place and in which the radioactive ^{8}B beam is produced. Two superconducting solenoids are used to collect, transport, analyze and focus the beam into a secondary chamber, where the secondary reaction takes place. For this experiment, the secondary target was a natural Ni sheet of 2.22 mg/cm^{2}, of 8.9 cm $\times$ 8.9 cm, which was bombarded with the ^{8}B beam at three different mean fusion energies in the center of mass reference frame (19.9, 21.9 and 23.8 MeV).

The experimental arrangement is shown in Fig. 3. Four ΔE-E silicon surface-barrier telescopes (A, B, C and D) were placed at backward angles (112.5°, 127.5°, 142.5° and 157.5°) to detect the evaporated protons from the fused system, while two detectors (E and F) were placed at forward angles to monitor the beam. Before starting the experiment, another detector (G) was placed at the secondary target position to characterize the beam. The beam intensity was reduced by a factor of 10^{3} during the characterization and the detector was taken away during the real experiment. A sample of the beam spectrum for the highest experimental energy is shown in Fig. 4, where the main beam components (^{6}Li, ^{7}Be and ^{8}B) are clearly separated by time-of-flight. A small contamination due to the ^{7}Li and ^{3}He components of the beam might be present in the proton yield associated to ^{8}B since the corresponding times-of-flight are similar. In contrast to ^{8}B, the ^{7}Li beam energies are always below the corresponding Coulomb barrier. This, combined with the fact that proton multiplicities for ^{7}Li + ^{58}Ni are about a factor of two lower than the ones for the ^{8}B beam, leads to a negligible effect on the proton yields. For ^{3}He, however, there was an effect on the proton yields of about 10% so it was necessary to consider this contribution. The respective correction was only considered for the highest energy, since for the other two energies the ^{3}He could be separated from ^{8}B by time of flight.

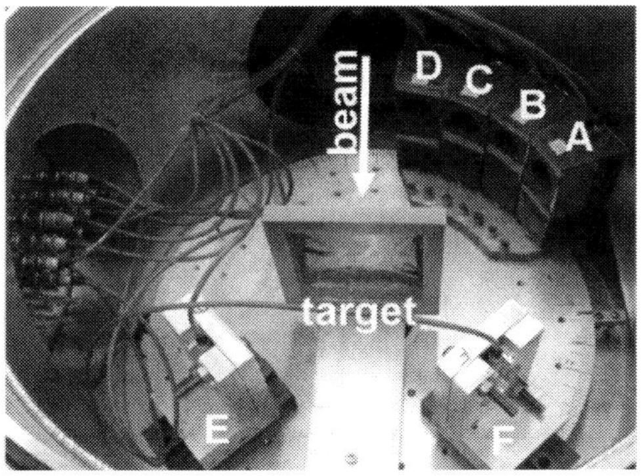

Figure 3. Experimental arrangement.

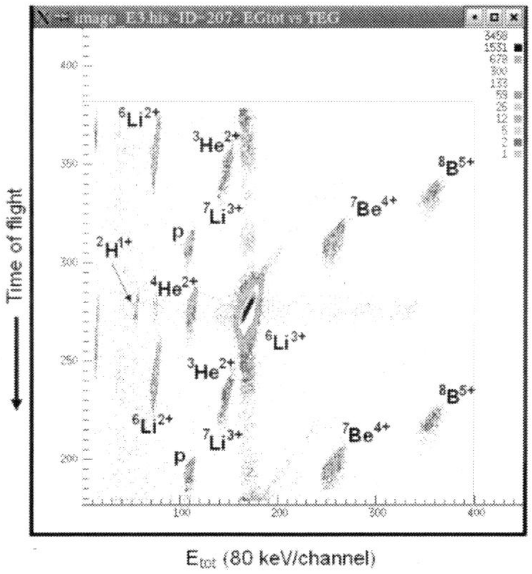

Figure 4. Beam composition.

4. Results and discussion

The results for the corresponding fusion excitation function for the three energies are presented in Fig. 5. Proton multiplicities calculated with the code PACE [22] were used to estimate these fusion cross sections. The solid line in this figure is a one-dimensional barrier penetration model (BPM) calculation and it is shown for comparison purposes. The corresponding barrier parameters ($V_b = 20.8$ MeV, $\hbar\omega = 5.53$ MeV and $R_b = 8.75$ fm) represent the typical barrier expected from the systematics for normal systems, which by comparison with the experimental data indicates a large fusion enhancement. This result seems to indicate an effect of the proton-halo of the ^{8}B nucleus.

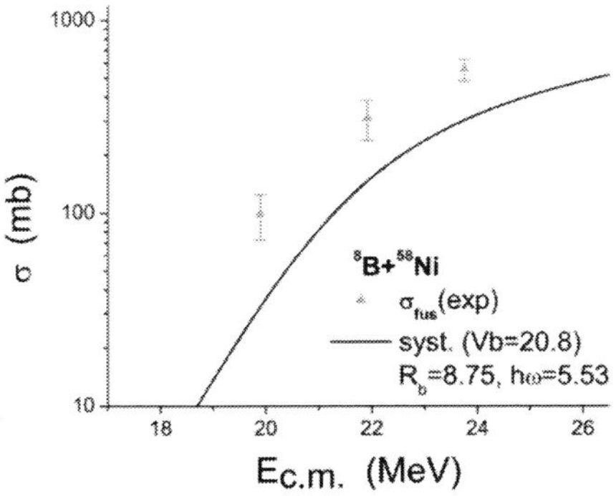

Figure 5. Fusion excitation function and the BPM calculation.

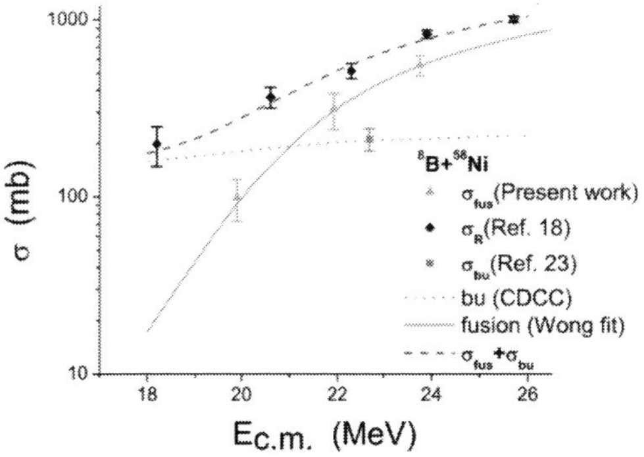

Figure 6. Fusion excitation function, total reaction and breakup cross sections for ^{8}B+^{58}Ni. The various curves represent calculations that are discussed in the text.

In Fig. 6, the solid line is a three parameter fit of the data with Wong's function [20]. An integrated breakup cross section from Ref. [1] is shown by the square and a CDCC calculation, performed for the ^{8}B breakup channel [23], is represented by the dotted line. Also displayed in this figure is the experimental total reaction cross section (black diamonds), reported for the ^{8}B+^{58}Ni system in Ref. [18]. As for the most important reaction channels associated to the present system, it can be seen that the sum of the CDCC calculations for the breakup plus the fusion cross sections reported in the present work (sum shown by the dashed line), seem to saturate the total reaction cross section. We conclude then that other possible reaction channels such as the proton transfer channel should be, at most, very weak. This conclusion was in fact anticipated both from comparison with the ^{7}Be core in Ref. [18] and from respective CDCC calculations in Ref. [23].

5. Conclusions

Evaporation protons from the ^{8}B+^{58}Ni system were measured at three near barrier energies. The proton multiplicities calculated with the code PACE were used to estimate the respective fusion cross sections. The resulting excitation function shows a large enhancement with respect to expectations for normal systems, an effect that is most probably related to the proton-halo nature of ^{8}B. By considering the results for the different reaction channels for this system, evidence is presented indicating that the breakup and the fusion channels saturate the total reaction cross section. This is consistent with previous estimations [18, 23] about the proton transfer channel being weak or negligible.

Acknowledgments

We would like to acknowledge the hospitality of all personnel at the Notre Dame Nuclear Structure Laboratory. This work has been partially supported by CONACYT, México, and by the US NSF.

References

[1] Guimarães V *et al.* 2000 *Phys. Rev. Lett.* **84** 1862
[2] Kolata J J *et al.* 2001 *Phys. Rev.* C **63** 024616
[3] Trache L, Carstoiu F, Gagliardi C A and Tribble R E 2001 *Phys. Rev. Lett.* **87** 271102
[4] Cortina-Gil D *et al.* 2002 *Phys. Lett.* B **529** 36
[5] Cortina-Gil D *et al.* 2003 *Nucl.Phys.* A **720** 3
[6] Majumdar R N 2003 *J.Phys.Soc.Jpn.* **72** 3087
[7] Hussein M S, Lichtenthäler R, Nunes F M and Thompson I J *et al.* 2006 *Phys. Lett.* B **640** 91
[8] Sumikama T, Nagatomo T, Ogura M, Iwakoshi T, Nakashima Y, Fujiwara H, Matsuta K, Minamisono T, Fukuda M and Mihara M *et al.* 2006 *Phys. Rev.* C **74** 024327
[9] Bertulani C A 2007 *J. Phys. G: Nucl. Part. Phys.* **34** 315
[10] Karataglidis S and Amos K 2007 *Phys. Lett.* B **650** 148
[11] Kolata J J *et al.* 1998 *Phys. Rev. Lett.* **81** 4580
[12] Aguilera E F *et al.* 2000 *Phys. Rev. Lett.* **84** 5058
[13] Aguilera E F *et al.* 2001 *Phys. Rev.* C **63** 061603(R)
[14] DiPietro A *et al.* 2004 *Phys. Rev.* C **69** 044613
[15] Navin A *et al.* 2004 *Phys. Rev.* C **70** 044601
[16] Milin M *et al.* 2004 *Nucl.Phys.* A **730** 285
[17] Benjamim E A *et al.* 2007 *Phys. Lett.* B **647** 30
[18] Aguilera E F *et al.* 2009 *Phys. Rev.* C **79** 021601(R)
[19] Kolata J J and Aguilera E F 2009 *Phys. Rev.* C **79** 027603
[20] Wong C Y 1973 *Phys. Rev. Lett.* **31** 766
[21] Lee M Y *et al.* 1999 *Nucl. Instrum. Methods Phys. Res.* A **422** 536
[22] Gavron A 1980 *Phys. Rev.* C **21** 230
[23] Belyaeva T L, Aguilera E F, Martinez-Quiroz E, Moro A M and Kolata J J 2009 *Phys. Rev.* C **80** 064617

Journal of Physics: Conference Series **322** (2011) 012008 doi:10.1088/1742-6596/322/1/012008

Woods-Saxon and São Paulo optical model calculations of the threshold anomaly of the ^{6,7}Li+^{28}Si systems near Coulomb barrier energies

A Gómez Camacho[1], P R S Gomes[2] and J Lubian[2]

[1] Departamento de Aceleradores, Instituto Nacional de Investigaciones Nucleares, Apartado Postal 18-1027, C.P. 11801, México, D.F., Mexico

[2] Instituto de Fisica, Universidade Federal Fluminense, Avenida Litoranea s/n, Gragoatá, Niterói, RJ, cep 24210-340, Brazil

E-mail: arturo.gomez@inin.gob.mx

Abstract. Two types of optical model interactions, namely the double-folding density dependent São Paulo potential (SPP) and the Woods-Saxon potential (WSP) are used to study the appearance of the Theshold Anomaly (TA) or the Breakup Threshold Anomaly (BTA) for the nuclear systems ^{6,7}Li+^{28}Si for energies around the Coulomb barrier. In the first case, the presence of the anomaly is determined from the energy dependence of the normalization parameters N_R and N_I as found from fittings to elastic scattering angular distributions. In the second case, the presence of the anomaly is investigated by the energy variation of the fusion and direct reaction parts of the Woods-Saxon potential. In this case, the parameters of the fusion and direct reaction potentials are calculated by a simultaneous analysis of elastic scattering, fusion and total reaction cross sections. It is found in both cases that the BTA appears for the ^{6}Li+^{28}Si reaction. As for ^{7}Li+^{28}Si, the calculation with the SPP shows that the usual TA appears but the dispersion relation is not satisfied, however the WSP indicates that indeed the BTA shows up.

1. Introduction

The presence of the Threshold Anomaly (TA), *i.e.*, the distinctive energy dependence behavior of the real and imaginary parts of the optical potential obtained from elastic scattering fits around the Coulomb barrier, has been well established in reactions with stable tightly bound nuclei. The characteristic behavior of the imaginary part of the optical potential is such that its strength sharply drops as the energy decreases towards the barrier energy. On the other hand, the real part presents a strong localized peak just around the barrier[1]. This phenomenon is understood in terms of the strong coupling between the elastic channel to other reaction channels that produces an additional attractive polarization potential $\Delta V(E)$. The absolute value of this attractive real polarization potential increases the strength of the already attractive nuclear potential V_0 which in turn lowers the fusion barrier and consequently produces an enhancement of the fusion cross sections at energies below the Coulomb barrier. This energy dependence of the real potential is expressed so by,

$$V(E) = V_0 + \Delta V(E), \tag{1}$$

where V_0 is the real nuclear potential at high energies and $\Delta V(E)$ the polarization potential. As the energy approaches the barrier, the imaginary potential $W(E)$ decreases as a consequence of the closing of reaction channels. That is, fusion and non-elastic peripheral reactions gradually become less important as coupling to the corresponding reaction channels decrease. The dispersion relation connects the energy variation of the real and imaginary parts of the polarization potential, this relation is expressed by the principal integral value[2],

$$\Delta V(E) = \frac{\mathcal{P}}{\pi} \int_0^\infty \frac{W(E')}{E' - E} dE'. \tag{2}$$

It is clear that any strong change in $W(E)$ must be accompanied by a localized strong variation in $\Delta V(E)$ as occurs in the TA. Recently, several studies have been dedicated to the case when weakly bound are involved instead of tightly bound nuclei. In this case, the energy dependence of the real $V(E)$ and imaginary $W(E)$ potentials around V_B is quite different. Weakly bound projectiles with a low threshold energies present strong couplings between the breakup and elastic scattering channels. That is, around and even for energies below the Coulomb barrier, reactions with weakly bound nuclei show strong direct reaction yields, particularly breakup or transfer cross sections. In such a situation, the imaginary part $W(E)$ of the optical potential can not sharply decrease, as it does for tightly bound nuclei. According to Eq.(2), the real polarization potential $\Delta V(E)$ does not contribute to increase the strength of the attractive nuclear potential, but instead has a repulsive character. This in turn increases the fusion barrier instead of lowering it and thus hinders fusion cross sections. This new phenomenon has been named the Breakup Threshold Anomaly (BTA) in contraposition to the usual Threshold Anomaly (TA)[3].

The BTA has been studied in a large number of nuclear systems involving weakly bound projectiles with a variety of medium and large mass targets. For instance the weakly bound projectile ^{9}Be with targets such as ^{27}Al[4], ^{64}Zn[5, 6, 7, 8], ^{144}Sm[9, 10] and with ^{208}Pb,^{209}Bi [11, 12]. Projectiles such as ^{6}Li and ^{7}Li have been investigated in reactions with ^{208}Pb[3, 13], ^{27}Al[14, 15, 16], ^{144}Sm[17], ^{58,64}Ni [18, 19], ^{59}Co [20], ^{90}Zr [21] , ^{138}Ba [6, 22, 23] , ^{28}Si [24]. As a general trend, the BTA is observed in all systems involving ^{6}Li, but for ^{7}Li its presence is observed for some systems and not for others. This may probably be due to the competition between the attractive polarization potential due to the bound excited state of ^{7}Li [23] and the repulsive breakup polarization potential [25]. The BTA has also been observed is some neutron and proton radioactive halo nuclei such as, ^{6}He with ^{209}Bi[26], and ^{8}B with ^{58}Ni[19, 27].

In this paper, we consider the extensive experimental work for elastic scattering, transfer and breakup cross section for the systems ^{6,7}Li + ^{28}Si done by A. Pakou and collaborators[24, 28, 29, 30, 31, 32, 33, 34, 35]. For these nuclear systems, their measurements show that transfer cross sections become higher than the breakup cross sections at near barrier energies. They argue that transfer channels can also be important at low energies for some weakly bound systems, and therefore those channels may be the main responsible for the non-decreasing of the imaginary potential at near barrier energies. We intend to study the presence of the TA or BTA for the ^{6,7}Li + ^{28}Si, by using two different polarization potentials, so that the conclusions reached are potential independent. The parameter-free non-local São Paulo potential (SPP)[36] has been extensively used to describe elastic scattering of nuclei such ^{6,7}Li and ^{9}Be with medium and heavy mass targets while Woods-Saxon potentials (WSP) with volume and surface parts have also been applied to the simultaneous description of fusion and elastic scattering data of various systems[9, 19, 37]. In the latter case, the volume potential is determined so as to fit fusion cross section data, whereas the surface part is responsible for direct reactions, including breakup and transfer processes. In the search of the volume part of the Woods-Saxon potential, fusion cross sections measurements of M. Sinha *et al.*, will be considered[38, 39, 40]. In the next section, a brief description of the SPP and WSP will be given. Section III will be concerned with the

Journal of Physics: Conference Series **322** (2011) 012008 doi:10.1088/1742-6596/322/1/012008

results for the Threshold Anomaly for the systems studied and finally, section IV is dedicated to a brief summary and conclusions.

2. Brief description of the theoretical framework

The Hamiltonian H for the nuclear system is of the form,

$$H_a = T_a + \mathcal{V}_a, \tag{3}$$

where the distorted wave $\chi_a^{(+)}$in the incident elastic channel a satisfies the expression,

$$(T_a + \mathcal{V}_a)\chi_a^{(+)} = E_a\chi_a^{(+)}. \tag{4}$$

The potential $\mathcal{V}_a$ is defined by,

$$\mathcal{V}(r, E) = V_{Coul}(r) - U_N(r, E), \tag{5}$$

where the subindex a has been suppressed. Here, $V_{Coul}(r)$ is the Coulomb potential and $U_N(r, E)$ is the energy dependent nuclear potential which is the sum of a bare potential with a slow energy dependence, plus the dynamic polarization potential which may have rapid variations at near barrier energies.

For the dynamical polarization potential, the best physical approach might be to use a double-folding potential for the real part with realistic matter densities and any reasonable approach for the imaginary part. This imaginary part can be constructed with the same radial dependence as the real part (double-folding potential) with a strength coefficient obtained at energies well above the barrier. However, when one studies the optical potential at near barrier energies, strength coefficients are needed to account for dynamic polarizations. A normalized version of the São Paulo optical potential [36, 41] reads as,

$$U_{SP}(r, E) = [N_R(E) + iN_I(E)]F(r, E), \tag{6}$$

where $F(r, E)$ is the double-folding potential, *i.e.*,

$$F(r, E) = V_F(r)e^{-4v^2/c^2}. \tag{7}$$

Here c is the speed of light, v the relative velocity of the colliding ions given by,

$$v^2(r, E) = \frac{2}{\mu}\left[E - V_{Coul}(r) - U_N(r, E)\right], \tag{8}$$

and $V_F(r)$ is the folding potential given in terms of the matter densities of the ions *i.e.*,

$$V_F(r) = \int \rho(\mathbf{r_1})\rho(\mathbf{r_2})V_0\delta(\mathbf{r} - \mathbf{r_1}+\mathbf{r_2})d\mathbf{r_1}\mathbf{dr_2}, \tag{9}$$

where a zero-range approximation is assumed with strength $V_0 = -456$ MeV-fm^3. In order to provide a parameter-free description of the interaction an extensive systematics of nuclear densities has been proposed by L.C. Chamon[41]. It has been found within the two-parameter Fermi distribution approach to describe the densities that the radii of the distributions are given by,

$$R_0 = 1.31A^{1/3} - 0.84, \tag{10}$$

where A is the number of nucleons in the nucleus. The systematics also shows that the matter diffuseness a of the density distributions has an average value $a = 0.56$ fm throughout the periodic table. The normalization of the double-folding potential e^{-4v^2/c^2}arising from the Pauli non-locality due to exchange of the nucleons vanishes at near barrier energies, and therefore the São Paulo potential becomes an usual double-folding potential.

The energy dependent normalization parameters $N_R(E)$ and $N_I(E)$ of Eq. (6) take into account the effects of the dynamic polarization potentials (DPP) arising from direct channel couplings. The real and imaginary parts of the DPPs should be related by the dispersion relation[3], *i.e.*,

$$N_R(E) = N_{R0} + \Delta N_R(E), \tag{11}$$

while,

$$\Delta N_R(E) = \frac{1}{\pi}\mathcal{P}\int_0^\infty \frac{N_I(E')}{E'-E}dE'. \tag{12}$$

N_{R0} is energy independent and is determined at some high enough energy for which the values of $\Delta N_R(E)$ and N_I are known. The energy dependence of the parameters N_R and N_I are calculated by fitting the elastic scattering angular distributions.

In order to obtain model independent conclusions about the threshold anomaly, a combined use of different approaches for determining the energy dependence of the optical potential is desirable. For this reason, we also use Woods-Saxon shape polarization potentials to calculate the energy dependence of such nuclear polarization potential. In this case, the total interaction between projectile and target of Eq.(5) is

$$\mathcal{V}(r,E) = V_{Coul}(r) - V_0(r) - U_N(r,E), \tag{13}$$

where $V_0(r)$ is the energy independent average nuclear potential and $U_N(r,E)$ is the nuclear dynamic polarization potential given by,

$$U_N(r,E) = V(r,E) + iW(r,E). \tag{14}$$

The real and imaginary potentials $V(r,E)$ and $W(r,E)$ are split into a fusion and direct reaction parts. That is, $V(r,E) = V_F(r,E) + V_{DR}(r,E)$ and $W(r,E) = W_F(r,E) + W_{DR}(r,E)$. The fusion potentials account for nuclear fusion processes σ_F whereas direct reaction potentials for direct reaction cross sections $\sigma_{DR} = \sigma_R - \sigma_F$. The potentials V_F and W_F are assumed to have Woods-Saxon volume shapes with the same parameters while V_{DR} and W_{DR} have surface geometrical forms also with the same parameters. The parameters of the potentials are calculated by a simultaneous χ^2-analysis of elastic scattering and fusion cross section data. It should noticed that that real fusion and direct reaction potentials are connected by the dispersion relation as in Eq.(2),

$$\Delta V_{F,DR}(E) = \frac{1}{\pi}\mathcal{P}\int_0^\infty \frac{W_{F,DR}(r,E')}{(E'-E)}dE' \tag{15}$$

Fusion and direct reaction cross sections are calculated by the well known equation,

$$\sigma_i = (2/\hbar v) < \chi^{(+)}|W_i|\chi^{(+)} >, i = F, DR \tag{16}$$

being $\chi^{(+)}$ distorted wave the solution of Eq.(4). For the nuclear systems under study, namely $^{6,7}Li+^{28}Si$, the energy variation of the parameters $N_R(E)$ and $N_I(E)$ of the SPP is thus studied as the collision energy approaches the Coulomb barrier. These parameters are obtained by fitting the experimental elastic scattering angular distributions for each system. On the other hand, the same energy dependence of $V_{F,DR}(E)$ and $W_{F,DR}(E)$ of the Woods-Saxon potential is also searched from a simultaneous fitting of elastic scattering and fusion data. Conclusions about the presence of the usual Threshold Anomaly or the Breakup Threshold Anomaly can be drawn from the energy dependence of all these functions as will be shown in the next section.

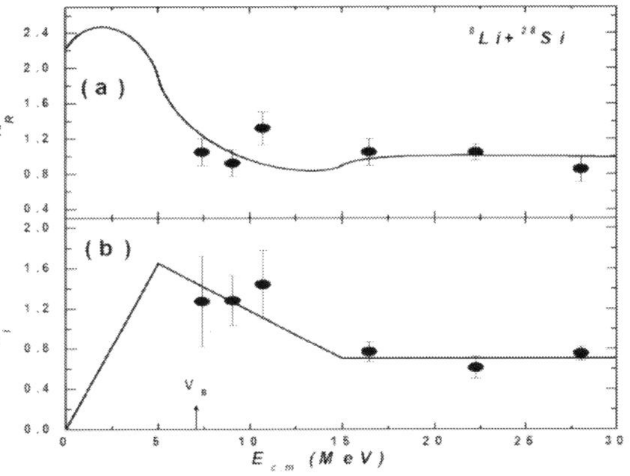

Figure 1. Real N_R and imaginary N_I parameters of the SPP as obtained from elastic scattering angular distributions and the dispersion relation calculation for ^{6}Li+^{28}Si.

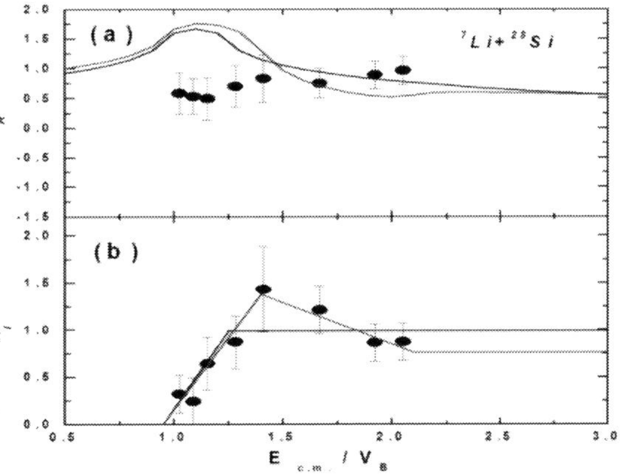

Figure 2. Same as Fig.1 but for ^{7}Li+^{28}Si.

3. Results and discussion

In the calculation of the parameters N_R and N_I of the double-folding interaction SPP and the fusion and direct reaction parameters of the Woods-Saxon potentials W_F and W_{DR} as described above, we consider the extensive elastic scattering angular distribution measurements by A. Pakou *et al.*,[28] for ^{6}Li+^{28}Si at the near barrier energies $E_{lab} = 7.5, 9, 11$ and 13 MeV. At higher energies, $E_{lab} = 20, 27$ and 30 MeV, the data from Refs.[42, 43, 44] are also used. For fusion cross sections σ_F, the recent sub- and above barrier measurements of M. Sinha *et al.*, are considered[38]. Total reaction cross sections σ_T are those from Refs.[29, 45]. As for the system ^{7}Li+^{28}Si, elastic scattering, fusion and total reaction cross section data are those of Refs. [24, 39, 40] and [29, 30], respectively. The dots in Figs.(1a), (1b), (2a) and (2b) show the results for N_R and N_I that best fit the elastic scattering angular distributions for both systems. The error bars shown are calculated by increasing the χ^2/N value in one unity, where N is the number of data points for each energy.

The linear fits shown for N_I are used to integrate the dispersion relation whose results are shown by the curves in the figures for N_R. These results show that for ^{6}Li+^{28}Si the BTA is present, since considering the error bars at the lowest energies an increasing trend is observed for $N_I(E)$ of Fig.(1b) as the collision energy is decreased towards the barrier V_B. On the other hand, the results for N_R of Fig.(1a), show a repulsive character as the energy decreases from high energies towards the barrier energy. Besides, it is observed that the result of the dispersion relation calculation shown by the curve in Fig.(1a) is close to the values obtained for N_R from the χ^2-analysis of elastic scattering distributions and so, the dispersion relation is satisfied. As for the ^{7}Li+^{28}Si system, the results for N_R and N_I are given in Figs.(2a) and (2b). It can be observed in Fig.(2b), the linear fits to the values obtained (dots) for N_I from the χ^2-fitting to the elastic distributions are more congruent with the usual TA. This is more evident precisely around the barrier $E_{c.m.}/V_B \approx 1$ where N_I decreases as the energy is decreased. The dots in Fig.(2a) are the values for N_R as obtained from the analysis of the data while the curves are the results of the integration of the dispersion relation that correspond to the linear fits of N_I. As observed, the results of the integration do not agree with the fitted values of N_R as the energy diminishes towards the barrier. As a matter of fact, a contrary character appears, that is the integration predicts an attractive polarization potential whereas the the fitting a repulsive character. The calculations of A. Pakou *et al.*,[24] for the system ^{7}Li+^{28}Si for the parameters N_R and N_I of the double-folding BDM3Y1 interaction show a similar behavior as ours.

We now consider the Woods-Saxon polarization potential with fusion and direct reaction parts. In the simultaneous χ^2-analysis of the fusion and elastic scattering angular distributions data, we fix the diffuseness and reduced radii of the fusion and direct reaction potentials for all energies as $a_F = 0.6$ fm, $r_F = 1.4$ fm, $a_{DR} = 0.6$ fm and $r_{DR} = 1.6$ fm for ^{6}Li+^{28}Si and $a_F = 0.6$ fm, $r_F = 1.4$ fm, $a_{DR} = 0.7$ fm and $r_{DR} = 1.6$ fm for ^{7}Li+^{28}Si. As described in Refs.[9, 19, 37], the real V_F and imaginary W_F parts of the Woods-Saxon fusion polarization potential are assumed to have the same geometric volume shape, with the same diffuseness and reduced radius. On the other hand, the corresponding direct reaction real and imaginary polarization potentials V_{DR} and W_{DR} have the same surface geometric shape, with the same diffuseness and reduced radius. So, only the strengths $V_F(E)$, $W_F(E)$, $V_{DR}(E)$ and $W_{DR}(E)$ are the variables to be obtained in the simultaneous analysis of the data. The results of the potential strengths for the system ^{6}Li+^{28}Si and ^{7}Li+^{28}Si are presented in Fig.(3) and Fig.(4). As seen in Figs.(3d) and (4d), as the collision energy approaches the corresponding barrier energy V_B, the fusion potentials $W_F(E)$ decrease even above the barrier, while the real counterparts $V_F(E)$ shown in Figs.(3c) and (4c) have an increasing behavior around the barrier. This is consistent with the expected behavior of the TA. Regarding the surface potential strengths $V_{DR}(E)$ and $W_{DR}(E)$, Figs.(3a) and (3b) for ^{6}Li+^{28}Si system give the results obtained from the simultaneous fitting (dots). It can be observed that these are compatible with the BTA. This is so since within the error bars, $W_{DR}(E)$ presents an increasing tendency as the collision energy approaches the barrier from higher energies, while $V_{DR}(E)$ presents a slight decrease. That is, the total strength of the real nuclear potential decreases resulting in a less attractive potential and therefore fusion reactions are hindered. As for the system ^{7}Li+^{28}Si, the values for $W_{DR}(E)$ (dots) shown in Fig.(4b) follow a slight increasing tendency as the energy decreases towards the barrier V_B. The linear fit of Fig.(4b) indicates that indeed an increasing behavior is adequate for $W_{DR}(E)$. This fact is an indicator that also the BTA is present for ^{7}Li+^{28}Si. On the other hand, it is true that for energies just around the barrier, a decreasing $W_{DR}(E)$ is present. The values for $V_{DR}(E)$ show a small repulsive behavior in the energy region where $W_{DR}(E)$ increases, being also an indicator that the BTA is present for ^{7}Li+^{28}Si.

The curves shown in Figs.(3a), (3c), (4a) and (4c) are the results of the dispersion relation integral corresponding to the linear forms of Figs.(3b), (3d), (4b) and (4d). Clearly, the results of the dispersion relation agree with the values obtained from the fitting of the data.

It is interesting to notice that since the diffuseness parameters a_F and a_{DR} and reduced radii r_F and r_{DR} of the potentials are kept constant for all collision energies for both nuclear systems, thus the energy behavior of the potential strengths shown by Figs.(3) and (4) is the same for

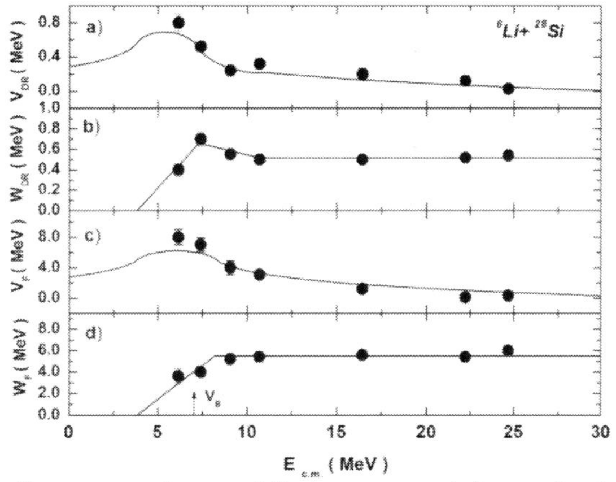

Figure 3. Strengths for direct reaction and fusion potentials as obtained from the simultaneous χ^2-analysis of elastic scattering, fusion and total reaction cross section data for ^{6}Li+^{28}Si.

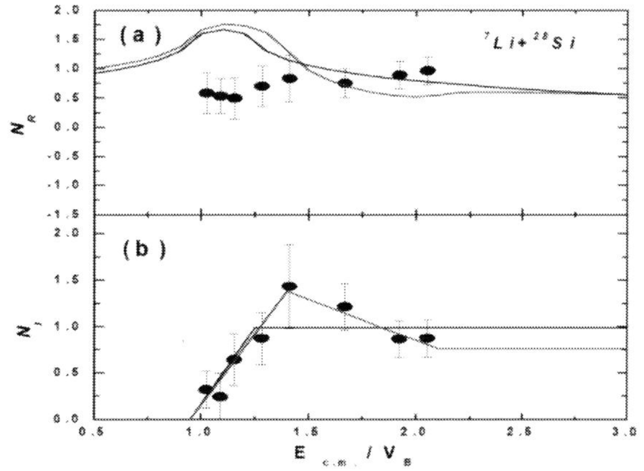

Figure 4. Same as Fig.3 but for ^{7}Li+^{28}Si.

the radial dependent volume and surface geometric functions $V_{DR}(E,r)$, $W_{DR}(E,r)$, $V_F(E,r)$ and $W_F(E,r)$, when these are evaluated at the strong absorption radius R_{sa}. This is illustrated in Fig.(5) where it also seen that $|W_F(E,R_{sa})| < |W_{DR}(E,R_{sa})|$ at all energies. That is, direct reactions dominate over fusion ones, particularly as the energy approaches the barrier energy V_B. As claimed by Pakou *et al.*, [28, 29, 24, 30, 31, 32, 33, 34, 35], direct reactions are not just breakup processes, but also transfer channels, which may even be more important than breakup at some energies for these systems.

The results of the simultaneous fits are given in Figs.(6) and (7) for fusion and reaction cross section data. The calculated values for fusion cross sections σ_F are shown in Fig. (6) for both systems. Reaction cross sections σ_T are shown in Fig. (7). For those energies where elastic scattering measurements are available but fusion data were not explicitly measured, we assumed those values obtained from a Wong fitting to the existing data[46]. Similarly, for those energies where reaction cross sections are not directly measured, we considered those values extracted from the corresponding elastic scattering data.

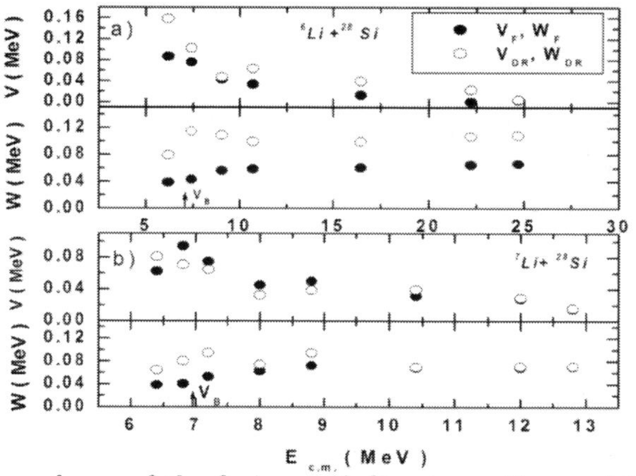

Figure 5. Energy dependence of the fusion and direct reaction real V and imaginary W parts of the polarization potentials at the corresponding strong absorption radius R_{sa} for a) ^{6}Li+^{28}Si and b) ^{7}Li+^{28}Si.

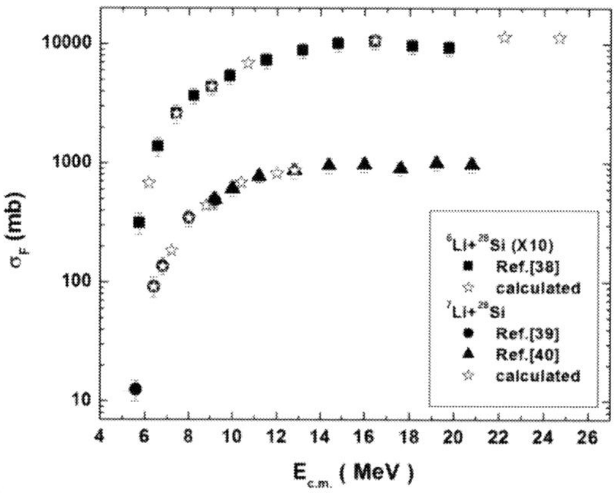

Figure 6. Calculated fusion cross section values as obtained from the simultaneous χ^2-analysis. To distinguish the systems, those values corresponding to ^{6}Li+^{28}Si have been multiplied by 10.

4. Summary and conclusions

Simultaneous fits of elastic scattering, fusion σ_F and total reaction cross sections σ_T for the systems ^{6}Li+^{28}Si and ^{7}Li+^{28}Si, at energies around and above the Coulomb barrier, have been presented. Two kinds of calculations have been performed: using a double-folding and Woods-Saxon potentials. In the calculations with the double folding São Paulo potential, the strengths of the real and imaginary potentials were varied to fit the data. For the ^{6}Li+^{28}Si system, clear indication of the presence of the BTA was observed, and the dispersion relation between real and imaginary potentials was satisfied. For the ^{7}Li+^{28}Si system, this evidence is not so clear, due to the competition between attractive and repulsive polarization potentials. Furthermore, the dispersion relation is not observed for this system. This fact seems to occur for some systems, when double folding potentials are used, and the explanation for that has to be investigated. In the calculations using Woods-Saxon potentials, volume and surface polarization potentials have been used to describe fusion and direct reaction cross sections, respectively. In the simultaneous

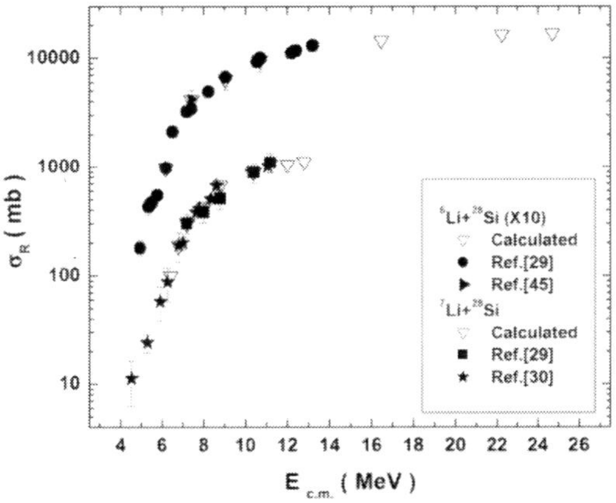

Figure 7. Calculated reaction cross section values as obtained from the simultaneous χ^2-analysis. To differenciate the systems, the experimental and calculated values corresponding to ^{7}Li+^{28}Si have been multiplied by 10.

χ^2-analysis, direct reaction cross section data are obtained from $\sigma_{DR} = \sigma_T - \sigma_F$. Also, in the present work, the energy dependence of the real and imaginary parts of the fusion potential and direct reaction potential have been determined. It has been found that the real and imaginary parts of the potentials, follow the dispersion relation for both systems. It is clearly observed that the fusion potential strengths, $V_F(E)$ and $W_F(E)$ behave according to the well known Threshold Anomaly. That is, $W_F(E)$ decreases as the energy approaches the barrier from higher energies and $V_F(E)$ has an increasing behavior for energies slightly above the barrier. On the other hand, the direct reaction potential strengths $W_{DR}(E)$ have a slight increasing tendency as the energy is lowered towards the barrier. The real $V_{DR}(E)$ manifests a small repulsive character in this same energy region. This conjugated effect for the energy dependence of the polarization potentials indicate that the systems ^{6}Li+^{28}Si and ^{7}Li+^{28}Si present the so-called Breakup Threshold Anomaly. This phenomenon is more clearly observed for the ^{6}Li+^{28}Si system than for ^{7}Li+^{28}Si. This fact can be understood because ^{6}Li has a smaller breakup threshold energy and no bound excited state, as compared with ^{7}Li. Since, the optical potential diffuseness and radial parameters of the fusion and direct reaction potentials are energy independent, the geometric functions $V_F(E,r)$, $W_F(E,r)$, $V_{DR}(E,r)$ and $W_{DR}(E,r)$ have the same energy dependence at the strong absorption radius R_{sa} as their strengths $V_F(E)$, $W_F(E)$, $V_{DR}(E)$ and $W_{DR}(E)$. It is thus observed that at R_{sa} , direct reaction potentials dominate over fusion ones at all energies, particularly as the energy approaches the barrier. This is in agreement with the fact that direct reactions, particularly breakup and/or transfer cross sections, are important and non-vanishing at energies below the barrier energy. The simultaneous χ^2-analysis of the data for both systems ^{6}Li+^{28}Si and ^{7}Li+^{28}Si are in close agreement to the data. Finally, in this work we have shown very clearly that the method of separating the optical potential into fusion and direct reaction potentials is a powerful tool to disentangle the opposite effects of production of attractive and repulsive polarization potentials by different reaction mechanisms. The two combining effects may lead to different net effects on the energy dependence of the total optical potential of different combinations of projectiles and targets.

Acknowledgments

P.R.S. G. and J. L. acknowledge the financial support from CNPq, PRONEX and FAPERJ.

Journal of Physics: Conference Series **322** (2011) 012008 doi:10.1088/1742-6596/322/1/012008

References

[1] Satchler G R 1987 *Nucl. Phys.* A **472** 591
[2] Mahaux C, Ngô H and Satchler G R 1986 *Nucl. Phys.* A **449** 354
[3] Hussein M S, Gomes P R S, Lubian J and Chamon L C 2006 *Phys. Rev.* C **73** 044610
[4] Gomes P R S *et al* 2004 *Phys. Rev.* C **70** 054605
[5] Gomes P R S *et al* 2005 *Phys. Rev.* C **71** 034608
[6] Gomes P R S *et al* 2005 *J. Phys. G* **31** S1669
[7] Moraes S B *et al* 2000 *Phys. Rev.* C **61** 064608
[8] Zadro M *et al* 2009 *Phys. Rev.* C **80** 064610
[9] Gómez Camacho A, Gomes P R S, Lubian J, Aguilera E F and Padrón I 2007 *Phys. Rev.* C **76** 044609
[10] Gomes P R S *et al* 2009 *Nucl. Phys.* A **828** 233
[11] Signorini C *et al* 2000 *Phys. Rev.* C **61** 061603(R)
[12] Woolliscroft R J *et al* 2003 *Phys. Rev.* C **68** 014611
[13] Keeley N *et al* 1994 *Nucl. Phys.* A **571** 326
[14] Figueira J M *et al* 2007 *Phys. Rev.* C **75** 017602
[15] Fernandez Niello J O *et al* 2007 *Nucl. Phys.* A **787** 484
[16] Figueira J M *et al* 2006 *Phys. Rev.* C **73** 054603
[17] Figueira J M *et al* 2010 *Phys. Rev.* C **81** 024613
[18] Biswas M *et al* 2008 *Nucl. Phys.* A **802** 67
[19] Gómez Camacho A *et al* 2010 *Nucl. Phys.* A **833** 156
[20] Souza F A *et al* 2007 *Phys. Rev.* C **75** 044601
[21] Kumawat H *et al* 2008 *Phys. Rev.* C **78** 044617
[22] Maciel A M M *et al* 1999 *Phys. Rev.* C **59** 2103
[23] Lubian J *et al* 2001 *Phys. Rev.* C **64** 027601
[24] Pakou A *et al* 2004 *Phys. Rev.* C **69** 054602
[25] Lubian J *et al* 2007 *Nucl. Phys.* A **791** 24
[26] Garcia A R *et al* 2007 *Phys. Rev.* C **76** 067603
[27] Lubian J *et al* 2009 *Phys. Rev.* C **79** 064605
[28] Pakou A *et al* 2003 *Phys. Lett.* B **556** 21
[29] Pakou A *et al* 2007 *Nucl. Phys.* A **784** 13
[30] Pakou A *et al* 2009 *Eur. Phys. J.* A **39** 187
[31] Pakou A *et al* 2003 *Phys. Rev. Lett.* **90** 202701
[32] Pakou A *et al* 2005 *Phys. Rev.* C **71** 064602
[33] Pakou A *et al* 2006 *Phys. Lett.* B **633** 691
[34] Pakou A *et al* 2007 *Phys. Rev.* C **76** 054601
[35] Pakou A 2008 *Phys. Rev.* C **78** 067601
[36] Chamon L C, Pereira D, Hussein M S, Candido Ribeiro M A and Galetti D 1997 *Phys. Rev. Lett.* **79** 5218
[37] Gómez Camacho A, Gomes P R S, Lubian J and Padrón I 2008 *Phys. Rev.* C **77** 054606
[38] Sinha M *et al* 2010 *Eur. Phys. J.* A **44** 403
[39] Sinha M *et al* 2007 *Phys. Rev.* C **76** 027603
[40] Sinha M *et al* 2008 *Phys. Rev.* C **78** 027601
[41] Chamon L C *et al* 2002 *Phys. Rev.* C **66** 014610
[42] Poling J E, Norbeck E and Carlson R R 1976 *Phys. Rev.* C **13** 648
[43] Cook J 1982 *Nucl. Phys.* A **375** 238
[44] Vineyard M F, Cook J and Kemper K W 1983 *Nucl. Phys.* A **405** 429
[45] Pakou A *et al* preprint NPL08_1 htpp://www.uoi.gr/physics/npl/preprints.htm.
[46] Wong C Y 1973 *Phys. Rev. Lett.* **31** 766

Journal of Physics: Conference Series **322** (2011) 012009 doi:10.1088/1742-6596/322/1/012009

Mechanism of heavy ion fusion to superheavy nuclei

Gurgen G. Adamian[1], Nikolai V Antonenko[1] and Werner Scheid[2]
[1]Joint Institute for Nuclear Research, 141980 Dubna, Russia,
[2]Institut für Theoretische Physik der Justus-Liebig-Universität, 35392 Giessen, Germany

E-mail: adamian@theor.jinr.ru, antonenk@theor.jinr.ru, werner.scheid@theo.physik.uni-giessen.de

Abstract. This article reviews different models for the description of fusion of heavy ions to superheavy nuclei by using adiabatic and diabatic potentials. The dynamics of fusion is basically different in the two types of models for fusion: In the adiabatic models the nuclei melt together, whereas in the diabatic models the nuclei transfer nucleons between each other up to the instant when the compound nucleus is formed. As final result we state that diabatic potentials seem more appropriate for the description of fusion of heavy nuclei than adiabatic potentials.

1. Introduction

Here we present a short discussion about the mechanism of heavy ion fusion. In the actual literature one can roughly divide the models in two different groups. The one type of models describes the fusion process as a melting of two nuclei and usually uses adiabatic internuclear potentials. The other type of models assumes that the nuclei stick together in a touching configuration and then exchange nucleons up to the time when the compound nucleus is formed. In this case, the nuclei first form a nuclear molecule, named dinuclear system by Volkov [1]. The potential between the nuclei is taken as a sudden one. Therefore, it sensitively depends on the choice of the internuclear potential which trajectory the nuclear system proceeds during the fusion process.

Fusion of two heavy nuclei to superheavy nuclei are reactions of the following kind:

cold fusion: $^{70}\mathrm{Zn} + {}^{208}\mathrm{Pb} \rightarrow {}^{277}\mathrm{Cn} + \mathrm{n},$
hot fusion: $^{48}\mathrm{Ca} + {}^{238}\mathrm{U} \rightarrow {}^{283}\mathrm{Cn} + 3\mathrm{n}.$

The dynamics of fusion is described by two main degrees of freedom, namely, by (a) the relative motion of the fusing nuclei, (b) the mass and charge transfer between the nuclei. Additional degrees of freedom are the deformations and orientations, the neck degree of freedom, internal excitations and other ones. For the mass and charge transfer we use the following coordinates

$$\eta = (A_1 - A_2)/(A_1 + A_2), \qquad \eta_Z = (Z_1 - Z_2)/(Z_1 + Z_2), \tag{1}$$

where A_1, A_2 and Z_1, Z_2 are the mass and charge numbers of the two clusters, respectively. In the case of $\eta = 0$ the nuclei with A_1 and $A_2 = A_1$ nucleons form a symmetric fragmentation; if $\eta = \pm 1$, the mass numbers A_2 or A_1, respectively, are zero which is interpreted as a fused system. So, the coordinate η varies in the interval (1, -1) and includes also the fused system.

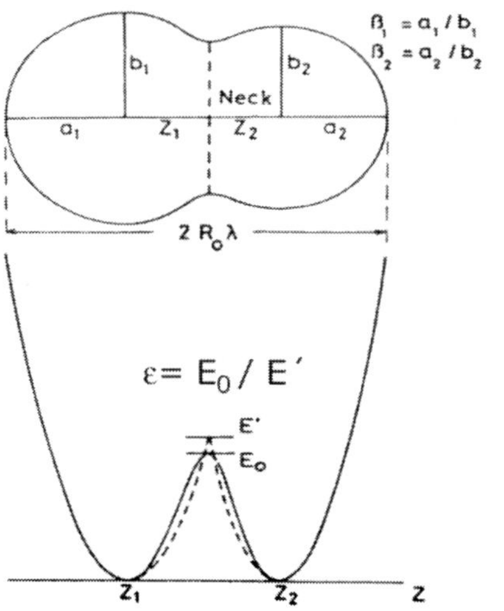

Figure 1. The parameters of the two-center shell model. Here, $2R_0\lambda$ measures the length of the system. The deformation parameters are given by $\beta_i = a_i/b_i$ with $i = 1, 2$. The neck parameter is $\epsilon = E_0/E'$. R_0 denotes the radius of the spherical compound nucleus.

The potential energy determines the dynamics of the various models of the fusion process. It can be calculated with the Strutinsky method where the potentials also contain shell effects:

$$U = U_{liquid\,drop} + \delta U_{shell}. \tag{2}$$

The shell effects in nucleus-nucleus collisions are usually determined by the two-center shell model of Maruhn and Greiner [2] (for the coordinates see figure 1).

The details of this short publication can be taken from the article "Clustering Effects within the Dinuclear Model", written by the authors for the second volume of the series "Clusters in Nuclei", edited by Christian Beck.

2. Adiabatic and diabatic potentials

For the description of the fusion process one uses adiabatic and diabatic potentials. Adiabatic potentials represent the minimum of energy for a set of given collective coordinates. In figure 2 we show the adiabatic potential for the ^{110}Pd+^{110}Pd collision as a function of the length $\ell = 2R_0\lambda$ of the nucleus system, where R_0 is the radius of the compound nucleus. The adiabatic potentials have an outside barrier, but in general no further inside barrier versus smaller λ ($\sim R$). They allow the formation of the compound system as a melting of two nuclei. Adiabatic potentials as a function of η are usually smallest for symmetric fragmentations ($\eta = 0$) and, therefore, give the largest fusion cross sections around $\eta = 0$ in contrast to the experimental data.

Diabatic potentials include the effects of relative motion. At every avoided crossing of the single-particle energies in the adiabatic two-center shell model the nucleons follow the so-called diabatic path, a process depending on the relative velocity of the nuclei and denoted as Landau-Zener effect for which the Pauli principle between the nuclei is responsible [3]. Diabatic two-center shell models can be treated with the maximum overlap method or the simpler method of maximum symmetry. The time-independent diabatic potential is then obtained on the basis of the adiabatic potential [4]

$$U_{diab}(R) = U_{adiab} + \sum_{\alpha} \varepsilon_{\alpha}^{diab}(R) n_{\alpha}^{diab}(R) - \varepsilon_{\alpha}^{adiab}(R) n_{\alpha}^{adiab}(R), \tag{3}$$

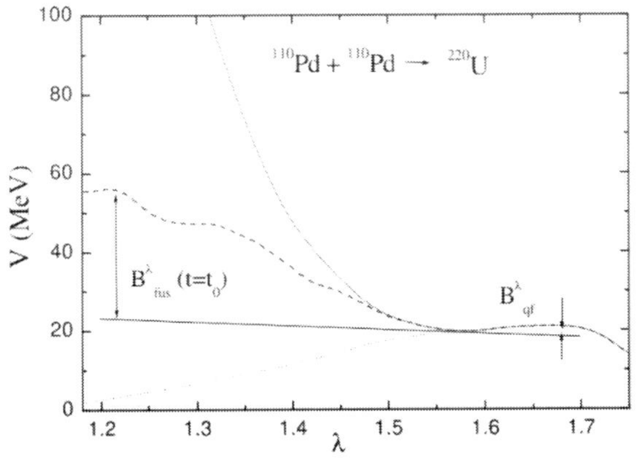

Figure 2. The diabatic (solid curve), the diabatic time-dependent (dashed curve) and the adiabatic (dotted curve) potentials for ^{110}Pd + ^{110}Pd as a function of λ.

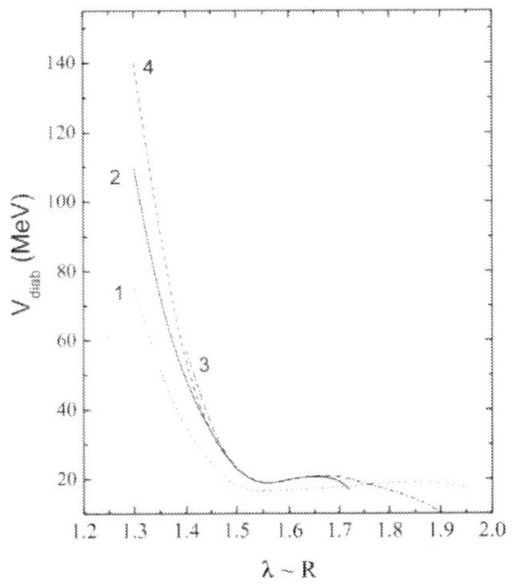

Figure 3. Diabatic potentials for the systems ^{100}Mo + ^{100}Mo (dotted line, curve 1) and ^{110}Pd + ^{110}Pd (solid line, curve 2). The phenomenological double folding potential for the system ^{110}Pd + ^{110}Pd is shown by the dashed-dotted line (curve 3). The discrepancy between this potential and the diabatic one becomes smaller if, by starting at the minimum of the pocket, the neck parameter ϵ is diminished by decreasing λ (dashed line, curve 4, for ^{110}Pd + ^{110}Pd).

where $\varepsilon_\alpha^{diab}$, $\varepsilon_\alpha^{adiab}$ are the single-particle energies of the diabatic and adiabatic two-center shell models and n_α^{diab}, n_α^{adiab} their occupation numbers, respectively. R is the internuclear coordinate.

Diabatic potentials are shown in figures 2 and 3. They are strongly repulsive and forbid fusion via the internuclear coordinate. As demonstrated in figure 3, they are similar to potentials calculated with the double folding method by using frozen densities (sudden potentials).

It is also possible to investigate the time-dependence of the transition from a diabatic potential to an adiabatic one. The evolution in time can be related to the characteristic relaxation time for the shape degrees of freedom of the system. The potential is calculated as [5]

$$U(\lambda,t) = U_{adiab}(\lambda) + \Delta U_{diab}(\lambda,t),$$
$$\Delta U_{diab}(\lambda,t) = \sum_\alpha \varepsilon_\alpha^{diab}(\lambda) n_\alpha^{diab}(\lambda,t) - \varepsilon_\alpha^{adiab}(\lambda) n_\alpha^{adiab}(\lambda). \tag{4}$$

Here, the occupation number $n_\alpha^{adiab}(\lambda)$ varies with λ according to a Fermi distribution with temperature $T(\lambda)$. The relaxation equations are given as

$$\frac{dn_\alpha^{diab}(\lambda,t)}{dt} = -\frac{1}{\tau(\lambda,t)}\left(n_\alpha^{diab}(\lambda,t) - n_\alpha^{adiab}(\lambda)\right), \tag{5}$$

where the relaxation time is chosen $\tau(\lambda,t) \simeq 5\times10^{-21}$s. The dashed curve in figure 2 is the time-dependent diabatic potential for ^{110}Pd + ^{110}Pd at a time $t_0 = 8\times10^{-21}$s which corresponds to the reaction time. One recognizes that at this time a barrier of about 55 MeV towards the inside remains which strongly suppresses the fusion along the relative coordinate.

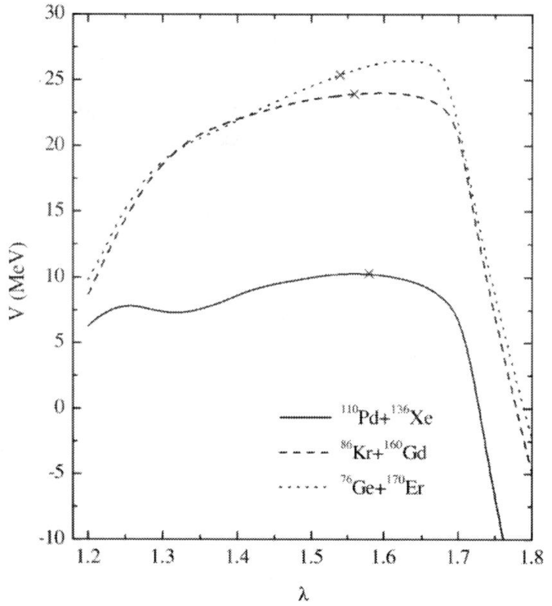

Figure 4. Adiabatic potential in different reactions leading to ^{246}Fm as a function of λ for a fixed $\varepsilon = 0.75$. The crosses denote the touching configurations.

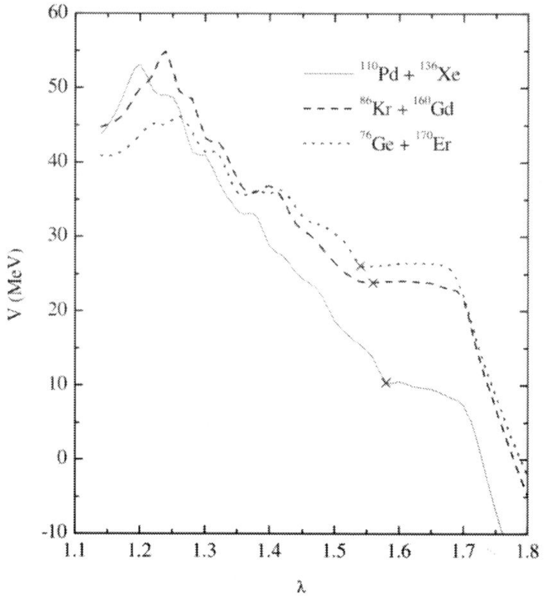

Figure 5. Dynamical diabatic potential. The notations are the same as in figure 4.

3. Comparison of models for fusion with adiabatic and diabatic potentials

3.1. Models using adiabatic potentials

These models minimize the potential energy and essentially apply an adiabatic dynamics in the internuclear distance. In this case the potential energy as a function of R and η allows a motion in R up to the fused compound nucleus. The adiabatic potential energy surface yields large probabilities for the fusion of equal projectile and target nuclei ($\eta = 0$) which contradicts the experimental data.

In figure 4 we show adiabatic potentials for collisions of nuclei forming the compound nucleus ^{246}Fm [6]. (The small crosses mean the touching points of the nuclei.) The corresponding fusion probabilities, given by the dotted curve in figure 6, result in the order of 0.1 (quite large). In contrast to the adiabatic potentials, the time-dependent diabatic potentials, drawn in figure 5 after the reaction, reveal barriers of about 50 MeV towards the inside [6]. With these potentials very small fusion probabilities are obtained. They are shown by the dashed curve in figure 6 which strongly deviates from the experimental values schematically presented by the solid curve in the same figure [6]. Therefore, we conclude that the fusion process can be described as a melting of nuclei along the relative coordinate with neither adiabatic nor diabatic potentials. These results force us to consider another fusion reaction mechanism by choosing a motion in the η coordinate.

3.2. Dinuclear system (DNS) concept

Models using diabatic potentials describe the fusion dynamics in the following steps: First, the nuclei are captured in the minimum of the adiabatic potential where they touch each other and form an excited nuclear molecule. This molecule consists of two individual nuclei. Then, a transfer of nucleons starts to change the ratio between the masses (also charges) of the nuclei up to the moment when the smaller nucleus lost all its nucleons to the larger one which becomes

Journal of Physics: Conference Series **322** (2011) 012009 doi:10.1088/1742-6596/322/1/012009

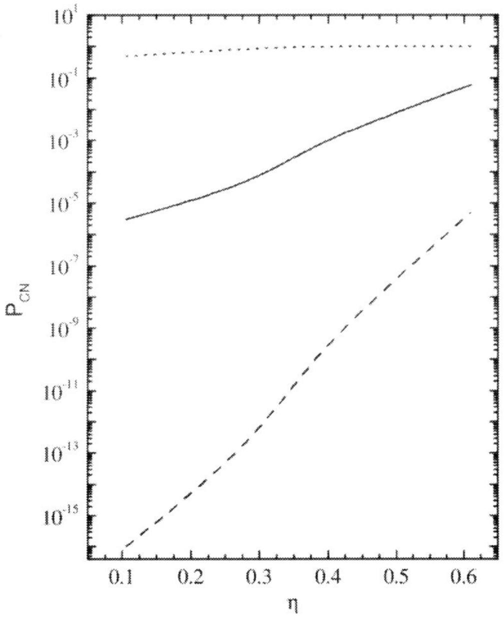

Figure 6. Fusion probability P_{CN} in the reactions leading to ^{246}Fm with excitation energy 30 MeV as a function of the mass asymmetry in the entrance channel. The result of the adiabatic treatment of the fusion in λ is presented by the dotted line. The upper limit of the fusion probability in λ in the dynamical diabatic treatment is presented by the dashed line. The fusion probability in the η channel with a closed fusion channel in λ is presented by the solid line.

the compound nucleus. The dynamics is a collective motion in the η (also η_Z) degree of freedom at a nearly fixed relative distance $R \approx 12$ fm which is roughly the sum of the nuclear radii. This concept is the dinuclear system model [1, 7] which treats the fusion as a transfer of nucleons at a fixed internuclear coordinate in contrast to the adiabatic models which explain the fusion as a melting of two nuclei along the relative coordinate at a nearly fixed η-value.

The potential at the touching point, in which the fusion process evolves by nucleon transfer, depends on the mass asymmetry coordinate η as shown in figure 7 and is called driving potential [8]. As one can recognize in figure 7, the driving potential is strongly influenced by the deformation of the fragments. It has a barrier, called inner barrier, for asymmetric fragmentations and is essentially diminished for symmetric systems which favor the fission of the dinuclear system in two (nearly) equal fragments, denoted as quasifission since no compound nucleus was existing before. The inner barrier determines the fusion probability in competition with the probability of quasifission. In figure 8 we present the probabilities for fusion in the cold reactions ^{A}X + ^{208}Pb and in the hot ones ^{48}Ca + ^{A}Y as a function of the Z_{CN}-value of the formed superheavy compound nucleus [9]. They can be calculated by master equations or by a statistical quantum mechanical method.

4. The motion of the neck

Let us consider the dynamics of the neck degree of freedom. The neck parameter ϵ is defined by the ratio of the actual barrier height E_0 to the barrier height E' of the corresponding two-center oscillator (see figure 1):

$$\epsilon = E_0/E' \tag{6}$$

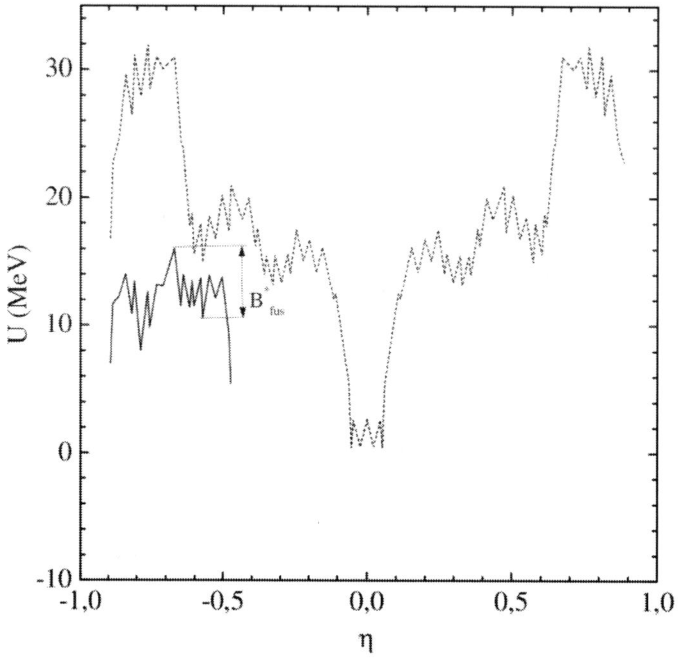

Figure 7. Potential energy of the dinuclear system in the reaction $^{54}Cr + ^{208}Pb \to ^{262}Sg$ reaction ($|\eta_i| = 0.59$) as a function of η for $J = 0$. Both curves are obtained with experimental binding energies. The dotted curve is calculated for spherical shapes of the nuclei, the solid curve for deformed shapes in pole-to-pole orientations.

The neck grows with decreasing ϵ. First, we made classic dynamical calculations with an adiabatic potential obtained with the Strutinsky method with shell effects calculated with the two-center shell model [10]. Figure 9 shows the potential energy surface as a function of the neck and length parameters ϵ and λ, respectively. The mass parameters were determined with the Werner-Wheeler approximation by assuming an incompressible and irrotational flow.

The fission-type valley is reached after a very short time of $3 - 4 \times 10^{-22}$ s at $\lambda \sim 1.68$ (see figure 9). Then oscillations in this valley occur if the kinetic energy is small. The characteristic time of all processes is of the order of 5×10^{-21} s. However, these calculations lead to a wrong dependence of the fusion probability on the isotopic composition and mass asymmetry in contrast to the experimental data. The neck grows too fast. There must exist a hindrance for the fast ϵ-increase and λ-motion towards the compound system. The essential hindrance which we found is a large, microscopically with the cranking model calculated mass parameter of the ϵ-motion. The main contributions to the mass tensor B^{cr}_{ij} result from the expression [10]

$$B^{cr}_{ij} \approx \hbar^2 \sum_{\alpha} \frac{f_\alpha}{\Gamma^2_\alpha} \frac{\partial \varepsilon_\alpha}{\partial q_i} \frac{\partial \varepsilon_\alpha}{\partial q_j} \qquad \text{with} \quad f_\alpha = -\frac{dn_\alpha}{d\varepsilon_\alpha}. \tag{7}$$

Here, ε_α and n_α are the two-center shell model single-particle eigenvalues and occupation numbers, respectively, and Γ_α are the widths of the decaying excited single-particle states. A much larger neck mass parameter than the one of the Werner-Wheeler approach results (cr=cranking, WW= Werner-Wheeler) [10]:

$$B^{cr}_{\lambda\lambda} = B^{WW}_{\lambda\lambda}, \quad B^{cr}_{\epsilon\epsilon} \approx 30 \times B^{WW}_{\epsilon\epsilon}, \quad B^{cr}_{\lambda\epsilon} \approx 0.35 \times B^{WW}_{\lambda\epsilon}. \tag{8}$$

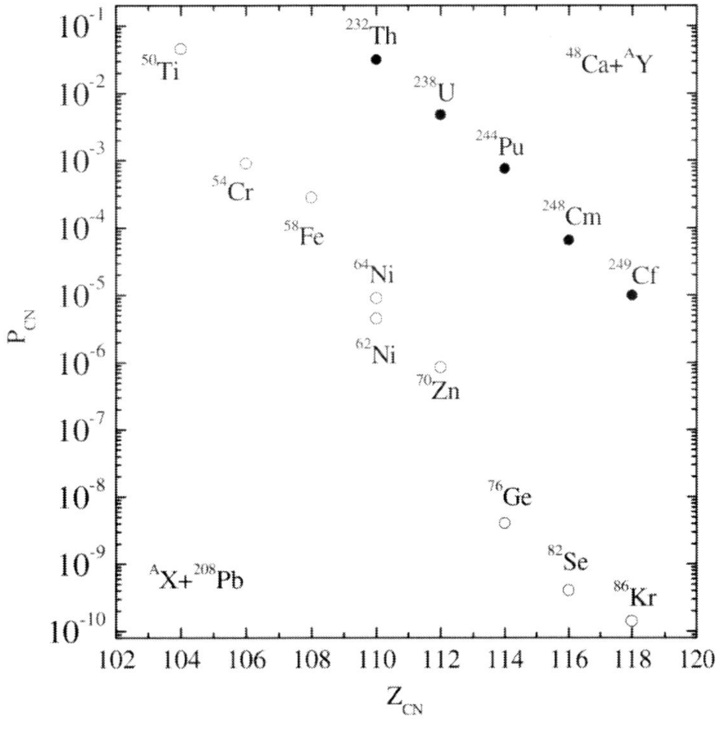

Figure 8. Calculated fusion probabilities P_{CN} for cold fusion reactions (^{A}X + ^{208}Pb) and hot fusion reactions (^{48}Ca + ^{A}Y) as a function of the charge number Z_{CN} of the superheavy compound nucleus.

As one can see in figure 10, the system stays a sufficiently long time near the entrance configuration. This justifies the dinuclear system model to a large extent. Then, mainly thermal fluctuations in the mass asymmetry degree of freedom are responsible for the process of fusion.

5. Summary

Fusion reactions for the production of superheavy nuclei are explained with adiabatic and diabatic potentials. The dynamics of fusion evolves differently in these potentials.

In adiabatic potentials the nuclei melt together along the internuclear distance. This process yields larger fusion cross sections for symmetric target and projectile combinations in contradiction to known experimental data.

Since diabatic potentials are strongly repulsive, the nuclei build up a dinuclear system of two touching nuclei and exchange nucleons up to the point when the compound nucleus is formed. This yields smaller fusion cross sections for symmetric fragmentations in agreement with the experimental data.

The formation of a larger neck is hindered by a large, microscopically calculated mass parameter for the neck degree of freedom. Therefore, the system hangs for some time in the initial touching configuration and can be interpreted as nuclear molecule or a dinuclear system. It can be analytically proved that a quantization of the kinetic energy tensor yields strongly repulsive potentials between the two approaching nuclei due to the interaction with excited states [11].

Several signatures for repulsive internuclear potentials between fusioning nuclei are found

Journal of Physics: Conference Series **322** (2011) 012009 doi:10.1088/1742-6596/322/1/012009

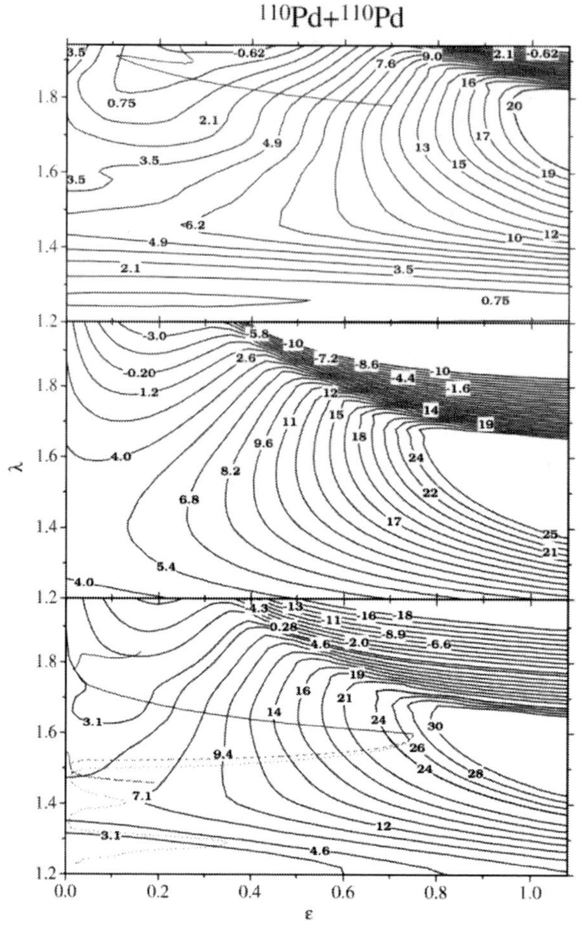

Figure 9. Potential energy surface (units MeV), calculated in the (λ, ϵ)-plane for the reaction ^{110}Pd + ^{110}Pd with shell corrections and $\beta_i = 1$ (lowest part), without shell corrections and $\beta_i = 1$ (middle part), and with shell corrections and $\beta_i = 1.2$ (upper part). The dynamical trajectories in the lowest part starting from the touching configurations and with initial kinetic energies 0, 40 and 60 MeV are presented by solid, dashed and dotted lines, respectively.

(*e.g.* Pauli principle). The calculated fusion and quasifission cross sections strongly point to the correctness of the diabatic description of the fusion process in comparison with the experimental data. This means that the ideas of the dinuclear system concept seem to give the correct description of the fusion process.

Acknowledgments
The authors thank Prof. V. V. Volkov for introducing them into the dinuclear system concept and Mrs. G. Scheid for a careful proofreading of this article.

References

[1] Volkov V V 1978 *Phys. Rep.* **44** 93; Volkov V V 1982 *Nuclear Reactions of Deep Inelastic Transfers* (Energoizdat: Moscow); Volkov V V 1986 *Izv. AN SSSR ser. fiz.* **50** 1879
[2] Maruhn J and Greiner W 1972 *Z. Phys.* **251** 431

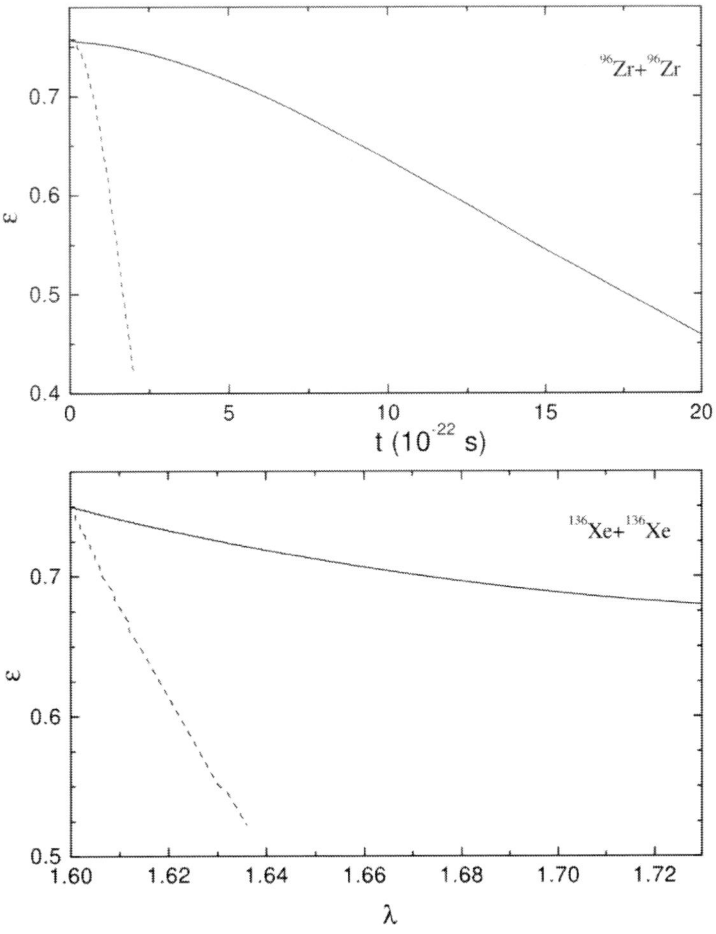

Figure 10. *Upper part:* Time-dependence of the neck parameter ϵ in the system ^{96}Zr + ^{96}Zr calculated with microscopical (solid curve) and Werner-Wheeler (dashed curve) mass parameters. *Lower part:* Trajectories in the (λ, ϵ)-plane calculated for the system ^{136}Xe + ^{136}Xe with microscopical (solid curve) and Werner-Wheeler (dashed curve) mass parameters. The end points of the solid and dashed curves are at time $t = 2 \times 10^{-21}$ s and $t = 2 \times 10^{-22}$ s, respectively.

[3] Adamian G G, Antonenko N V and Tchuvil'sky Yu M 1999 *Phys. Lett.* B **451** 289
[4] Diaz Torres A, Antonenko N V and Scheid W 1999 *Nucl. Phys.* A **652** 61
[5] Diaz Torres A, Adamian G G, Antonenko N V and Scheid W 2000 *Phys. Lett.* B **481** 228
[6] Adamian G G, Antonenko N V, Diaz Torres A and Scheid W 2000 *Nucl. Phys.* A **671** 233
[7] Adamian G G, Antonenko N V, Scheid W and Volkov V V 1997 *Nuov. Cim.* A **110** 1143
[8] Adamian G G, Antonenko N V, Scheid W and Volkov V V 1997 *Nucl. Phys.* A **627** 361
[9] Adamian G G, Antonenko N V and Scheid W 2000 *Nucl. Phys.* A **678** 24
[10] Adamian G G, Antonenko N V, Ivanova S P and Scheid W 1999 *Nucl. Phys.* A **646** 29
[11] Fink H J, Scheid W and Greiner W 1975 *J. Phys.* G **1** 685

Journal of Physics: Conference Series **322** (2011) 012010 doi:10.1088/1742-6596/322/1/012010

Phenomenological and semimicroscopic cluster models and their phase transitions

P R Fraser[1], H Yépez-Martínez[2], P O Hess[1] and L Parra Rodríguez[1]

[1] Instituto de Ciencias Nucleares, Universidad Nacional Autónoma de México, 04510 México, D.F., Mexico

[2] Universidad Autónoma de la Ciudad de México, 09790 México, D.F., Mexico

E-mail: paul.fraser@nucleares.unam.mx

Abstract. Two algebraic cluster models of nuclear structure which use the same Hamiltonian are explored: a Phenomenological Algebraic Cluster Model (PACM), and a Semi-microscopic Algebraic Cluster Model (SACM). The PACM does not incorporate the Pauli exclusion principle, while the SACM does. The Hamiltonian considered is an admixture of three dynamical symmetries; the $SU(3)$, $SO(4)$, and $SO(3)$, with weighting of each determined by parameters. The classical potential is constructed using coherent states, and the model Hilbert space of the SACM is constructed in such a way as to include all shell model states which correspond to the cluster structure of interest. Phase transitions and their orders are investigated for each model, and parameter phase diagrams are presented, wherein it is found that consideration of the Pauli principle has significant consequences. Also shown are fits of Hamiltonian parameters for a nucleus of 2 spherical clusters, when moving between the symmetry limits.

1. Introduction and formalism

To illustrate the significant effect of the Pauli principle in nuclear cluster studies, two algebraic nuclear cluster models are investigated: a Phenomenological Algebraic Cluster Model (PACM), and a Semimicroscopic Algebraic Cluster Model (SACM) [1, 2]. Both use a modified version of the vibron model [3] to describe relative motion, and the $SU(3)$ model of Elliott [4] for internal cluster structure. (The true vibron model describes clusters with the IBA [5, 6], and treats $\hbar\omega$ as a parameter, where here it is fixed by shell model considerations.) The difference between PACM and SACM is that the latter includes the Pauli exclusion principle for the combination of the nucleons of the two clusters.

It should be noted that here we deal not with nuclear molecules, such as the $\alpha + \alpha \rightarrow {}^8$Be system, which could be called 'strong' clusterisation. We deal rather with the 'weak' clusterisation of systems such as ^{16}O, whose states can be considered to have some overlap with experimental basis states of the clusters $\alpha+{}^{12}$C.

In both models, the relative motion degrees of freedom are oscillators in 3 dimensions, plus an additional auxiliary scalar boson, described by boson creation and annihilation operators:

$$\pi^\dagger_m,\ \pi_m,\ m = 0, \pm 1\ ;\ \sigma^\dagger, \sigma\ . \tag{1}$$

σ-bosons have no physical meaning, but define a cut-off for a conserved total number of bosons, $N = n_\pi + n_\sigma$.

In both PACM and SACM, the same Hamiltonian is used:

$$\hat{H} = xy\hat{H}_{SU(3)} + y(1-x)\hat{H}_{SO(4)} + (1-y)\hat{H}_{SO(3)} \ , \tag{2}$$

where x and y are parameters ranging from 0 to 1, determining the strength of influence of each symmetry. In their most general form, the individual components are

$$\hat{H}_{SU(3)} = \hbar\omega\hat{n}_\pi + a_C\hat{\mathcal{C}}_2(\lambda_C, \mu_C) + (a - b\triangle\hat{n}_\pi)\hat{\mathcal{C}}_2(\lambda, \mu) + \gamma\hat{L}^2 + t\hat{K}^2 \ , \tag{3}$$

$$\hat{H}_{SO(4)} = a_C\hat{\mathcal{C}}_2(\lambda_C, \mu_C) + a_C\hat{L_C}^2 + a_R^{(1)}\hat{L_R}^2 + \gamma\hat{L}^2 + \frac{c}{4}\left[\left(\hat{\pi}^\dagger\cdot\hat{\pi}^\dagger\right) - \left(\sigma^\dagger\right)^2\right]\left[(\hat{\pi}\cdot\hat{\pi}) - (\hat{\sigma})^2\right] \ , \tag{4}$$

and

$$\hat{H}_{SO(3)} = \hbar\omega\hat{n}_\pi + a_C\hat{\mathcal{C}}_2(\lambda_C, \mu_C) + a_C\hat{L_C}^2 + a_R^{(1)}\hat{L_R}^2 + \gamma\hat{L}^2 \ , \tag{5}$$

where $\triangle\hat{n}_\pi = \hat{n}_\pi - n_0$, with n_0 being the minimum number of quanta needed to satisfy the Pauli principle, as per the Wildermuth condition (see below). $n_0 = 0$ in the PACM. $\triangle\hat{n}_\pi$ is included as without it, $\hat{H} \sim \hat{n}_\pi - \hat{n}_\pi^2$, with the latter dominating at high n_π [7]. The $SO(3)$ limit is a special case of the $SU(3)$, with only the $\hbar\omega\hat{n}_\pi$ term plus couplings on the $SO(3)$ level. $\hat{K}$ is defined in [1, 2].

R represents relative motion and C denotes the internal structure of both clusters, the irreducible representations of the individual clusters being subsumed into irreducible representations of Elliott [4] in the relevant group chain, i.e. $SU_{C_1}(3) \otimes SU_{C_2}(3) \otimes SU_R(3) \supset SU_C(3) \otimes SU_R(3)$ [see Eq. (3) of [8] for full chain used]. That is, the cluster number $k : 1, 2$ is within a definite $SU_{C_k}(3)$ irrep (λ_k, μ_k), and both are coupled to (λ_C, μ_C). This is reflected on the operator level in Eq. (3) - (5). The generators of $SU_C(3)$ are the sum of the generators of $SU_{C_k}(3)$; for example, Q^C_{2m}, the quadrupole operator of $SU_C(3)$, is simply expressed $Q^C_{2m} = Q^{C_1}_{2m} + Q^{C_2}_{2m}$. Therefore, the second order Casimir operator is the sum of the Casimir operators of the individual clusters plus a coupling term.

Finally, (λ_C, μ_C) is coupled with the relative motion irrep, $(n_\pi, 0)$, yielding the total $SU(3)$ irrep (λ, μ). On the operator level, Eq. (3) - (5), this coupling is subsumed within the structure term. As an example of this, $\hat{L}^2 = \hat{L}_R^2 + \hat{L}_C^2 + 2\left(\hat{L}_R\cdot\hat{L}_C\right)$, reflecting the coupling of relative motion to the cluster irrep.

In the case of two spherical clusters, an example of which follows, $(\lambda_k, \mu_k) = (0, 0)$, and thus $(\lambda_C, \mu_C) = (0, 0)$, and so (λ, μ) is $(n_\pi, 0)$. All terms denoted with subscript C and fixed-body axis term, $t\hat{K}^2$, go to 0. For more details, see [1, 2].

In the study of phase transitions, it is necessary to define an intensive Hamiltonian, and one method is given in [9]. However, with a procedure based upon and constrained by the shell model, the situation is more subtle: it is not possible to divide the one-body interaction, $\hbar\omega\hat{n}_\pi$, by N, as this would lead, with increasing N, to a zero inter-shell distance. Thus, as outlined below, in this work the coherent state parameter is related to the intercluster distance, which results in an N-independent expectation value of $\hat{H}$. Thus, the result obtained is the same as in [9], but stemming from different arguments.

Consider again that we can write this one-body term as $\hbar\omega\left(n_0 + \triangle\hat{n}_\pi\right)$, and that $\hbar\omega n_0$ can be included in the zero-point energy, ε_0, of [9] (though this is yet to be substantiated). Using coherent states and comparing the geometric quadrupole operator with the algebraic version of the same, it is shown in detail in [8] that

$$\alpha \sim \frac{r - r_0}{\sqrt{N}} \ , \tag{6}$$

where r is the geometric inter-cluster distance and $r_0 \sim \sqrt{n_0}$ (showing that the inclusion of the minimal inclusion of the Pauli principle automatically generates a minimal distance). This

α is then used to form coherent states and thus a potential that does not go to infinity as N does. This can been seen in Eqs. (10) - (12) for PACM, where asymptotically, all N dependence vanishes (considering C to be a constant, which can be discarded, and c rescaled with N^{-1} dependency), and the same holds for SACM. A fuller account of this is in preparation.

This Hamiltonian is used to investigate phase transitions in clusterised nuclear matter, after defining a potential by forming diagonal matrix elements with coherent states [see Eq. (9) and (15)]. A first order phase transitions is defined as a line of transition between which of a potential's two wells is deepest. Crossing that line, the value of first derivative at the minima with respect to control parameters is not the same. They are said to be discontinuous. I.e.,

$$\frac{\partial V(\alpha_1)}{\partial P} \neq \frac{\partial V(\alpha_2)}{\partial P} , \tag{7}$$

where α_1 and α_2 are a minimum at the origin (spherical) with regards to one variable, and away from the origin (deformed) respectively. P represents control parameters.

A second order phase transition is a line of transition where, for our purposes, a deformed minimum in a potential becomes spherical. Crossing that line, the value of second derivative at the minima with respect to control parameters is not the same. I.e.,

$$\frac{\partial^2 V(\alpha_1)}{\partial P^2} \neq \frac{\partial^2 V(\alpha_2)}{\partial P^2} . \tag{8}$$

This idea generalises to higher order phase transitions [10, 11].

2. PACM

PACM makes no attempt to account for the Pauli principle, and its coherent states are

$$|\alpha\rangle = \frac{1}{\sqrt{N![1+(\alpha\cdot\alpha)]^N}} \left[\hat{\sigma}^\dagger + \left(\alpha\cdot\hat{\pi}^\dagger\right)\right]^N |0\rangle , \tag{9}$$

where α are variables representing a measure of inter-cluster distance. This coherent state, when applied to the Hamiltonian of Eq. (2), (3), (4) and (5), gives a potential of the form

$$V = -N(N-1)(N-2)(xyb)\left\{A\frac{\alpha^2}{(1+\alpha^2)} - B\frac{\alpha^4}{(1+\alpha^2)^2} + \frac{\alpha^6}{(1+\alpha^2)^3}\right\} + C , \tag{10}$$

where A, B and C are linearly independent functions of x, y, the Hamiltonian parameters and N,

$$A = -\frac{xy[\hbar\omega + (a-b)\{4+(\Gamma_1+\Gamma_2)\} - b\mathcal{C}_2(\lambda_C,\mu_C)] - (1-x)yc(N-1) + (1-y)\hbar\omega}{xyb(N-1)(N-2)} , \tag{11}$$

$$B = \frac{x[a-6b-b(\Gamma_1+\Gamma_2)] + (1-x)c}{xb(N-2)} \tag{12}$$

and

$$C = xya\mathcal{C}_2(\lambda_C,\mu_C) + \tfrac{1}{4}(1-x)ycN(N-1) , \tag{13}$$

The central bracket of this, dependent on α with parameters A and B, is sufficient to investigate phase transitions, as shown in Fig. 1. In Region I, both a spherical and a deformed minimum exist, and the global minimum is deformed. In Region II, both a spherical and a deformed minimum exist, and the global minimum is spherical. In Region III only a spherical minimum exists, and in IV only a deformed. Solid lines are phase transitions, the horizontal being second order and the curved first order. Dashed lines are changes of region without phase

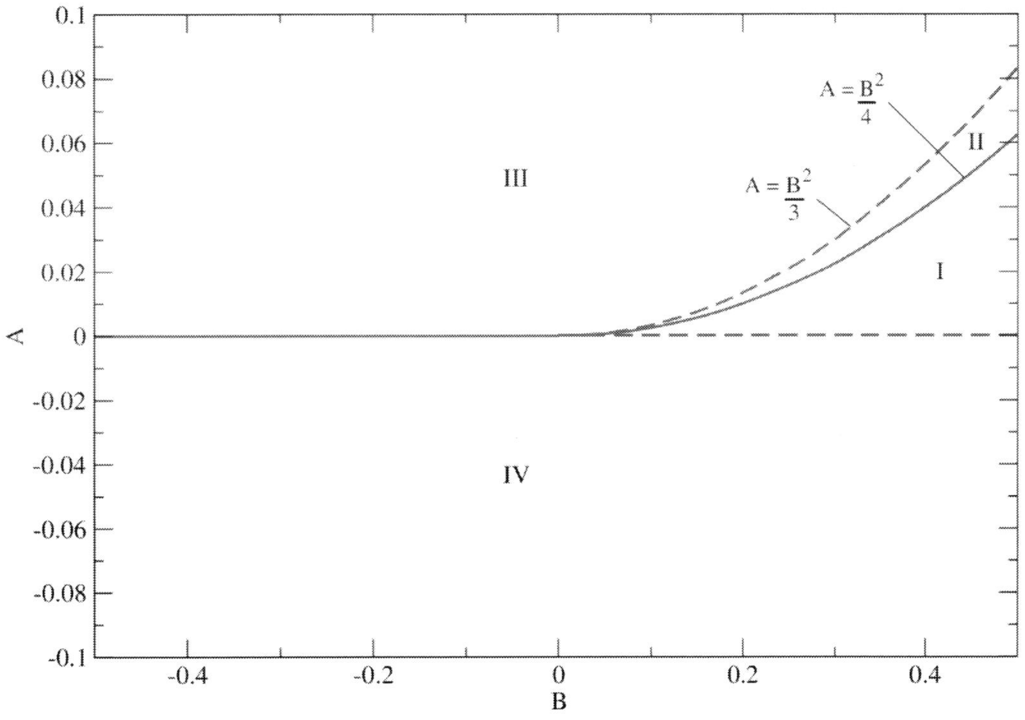

Figure 1. Phase space diagram for PACM. Explanatory notes regarding region numbers and line types in Section 2. This figure is independent of α.

transition. Without third order terms, $\triangle\hat{n}_\pi$, B is only defined for values less than zero, and thus only second order phase transitions exist, as in the model of [12].

Obtaining a fit to data for α+^{16}O , it was found that PACM is physically insufficient. When fixing parameters in the $SU(3)$ limit, considering the ground state band and restricting to even angular momentum the spectrum must start with $n_\pi = 0$, thus a 0^+ state, and then $n_\pi = 2$, so a 0^+ and a 2^+ state. However, the $SU(3)$ Hamiltonian contains γL^2, with γ always positive, which forces 0_2^+ to be lower in energy than the 2_1^+. This defies the known data $E(2_1^+) = 1.634$ MeV and $E(0_2^+) = 8.7$ MeV.

Also, the 2_1^+ state is of $1\hbar\omega$ excitation, where $\hbar\omega = 13.2$ MeV. Thus, the quadrupole-quadrupole interaction has to be unnaturally strong to locate this state at the experimental energy, 1.634 MeV. Such moves these higher-band states down to give the ground state band order 0_1^+, 2_1^+, 0_2^+, but the $\langle n_\pi\rangle$ of these states are then 20, 20 and 0 respectively, meaning what should be 0_1^+ is 0_2^+, and the first two states are at the cut-off, and thus don't converge.

Additionally, with different n_π values for states in the ground state band, there are different $SU(3)$ structures in said band, so it is not a rotation band.

As one cannot use the $SU(3)$ limit for this nucleus as microscopic models guide one to, the $SO(4)$ is next examined. The $SO(4)$ limit is different to the $SU(3)$ as, in the former, a state is a mixture of many basis states in $SU(3)$. Here data can be fit well. However, when the arbitrary boson cut-off, N, is increased, the structure of the states change, implying that no convergence is reached. This is illustrated by the result [13]:

$$\langle n_\pi\rangle = \frac{N-1}{2}\ . \tag{14}$$

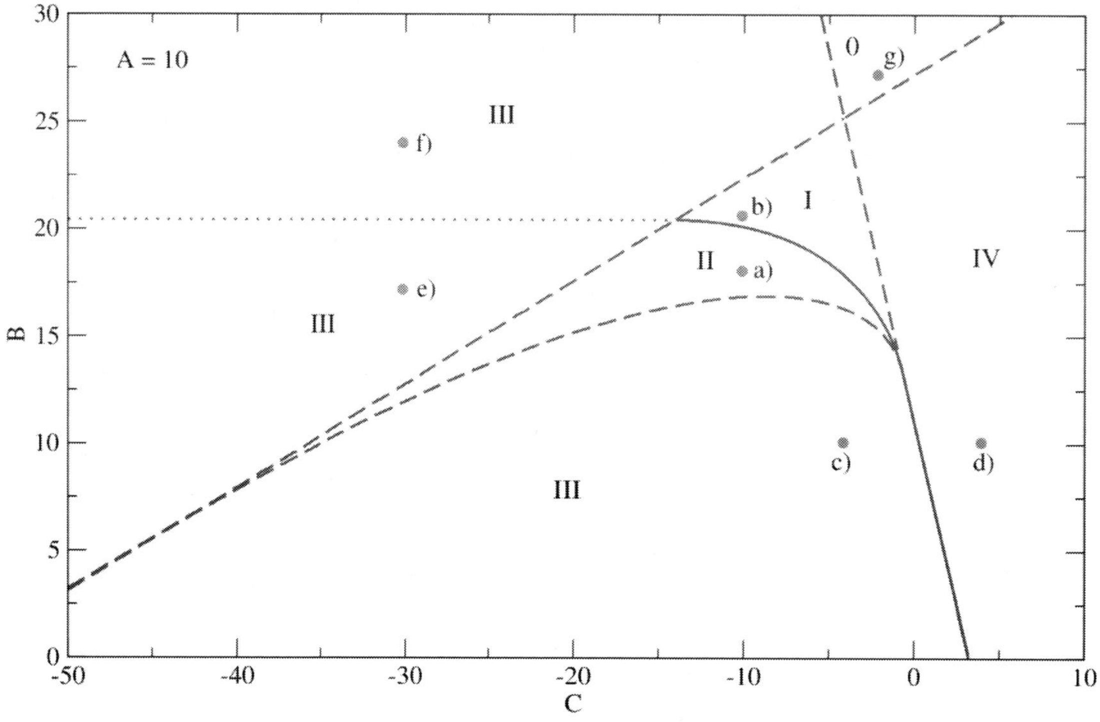

Figure 2. Phase space diagram for SACM. Explanatory notes regarding region numbers and line types are in the text. This figure is independent of α.

This implies high shell excitations, and assuming that the two clusters are moving in a shell model mean field, one cannot make $\hbar\omega$ a parameter to overcome this and maintain that framework. Note also that any result based on an arbitrary cut-off cannot be physical.

3. SACM

SACM accounts for the Pauli principle by only considering states not forbidden in shell model calculations. For simple cluster systems, the minimal condition for achieving this is the Wildermuth condition [14], fully discussed in [1, 2, 8]. Then, the model space is obtained by comparing all possible irreps (λ, μ), contained in $(\lambda_{C_1}, \mu_{C_1}) \otimes (\lambda_{C_2}, \mu_{C_2}) \otimes (n_\pi, 0)$, with those of the shell model. Only those irreps appearing in shell model results are retained. In this microscopic approach, the Pauli principle is applied or not, and cannot be 'tuned'. Note that $\hbar\omega$, which appears in the $SU(3)$ and $SO(3)$ Hamiltonians, must be a constant given by the shell model. It cannot, as in other models not guided by the shell model, be a parameter. The inclusion of n_0 from the Wildermuth condition yields the coherent state

$$|\alpha\rangle = \mathcal{N}_{N,n_0} \frac{N!}{(N+n_0)!} \frac{d^{n_0}}{d\gamma^{n_0}} \left[\hat{\sigma}^\dagger + \gamma \left(\alpha \cdot \hat{\pi}^\dagger \right) \right]^{N+n_0} |0\rangle \ , \tag{15}$$

where $N + n_0$ is now the total number of relative oscillation quanta, and the γ parameter has to be set equal to 1 after differentiation. This leads to the potential

$$V = -xyb \left(A\alpha^2 \frac{F_{11}(\alpha^2)}{F_{00}(\alpha^2)} - B\alpha^4 \frac{F_{22}(\alpha^2)}{F_{00}(\alpha^2)} + \alpha^6 \frac{F_{22}(\alpha^2)}{F_{00}(\alpha^2)} - C\alpha^2 \frac{F_{20}^{N-2}(\alpha^2)}{F_{00}(\alpha^2)} \right) + D \ , \tag{16}$$

Journal of Physics: Conference Series **322** (2011) 012010 doi:10.1088/1742-6596/322/1/012010

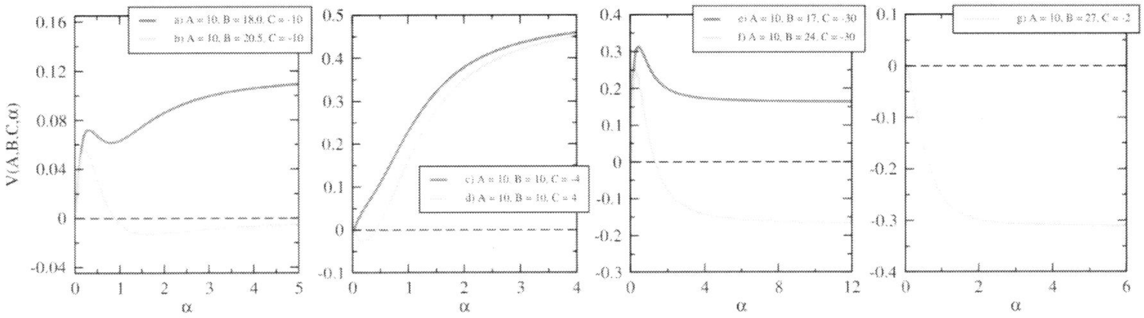

Figure 3. Graphs of SACM. Left to right: crossing 1st order phase transition, crossing 2nd order phase transition, crossing between types of Region III, region of no minimas.

where A, B, C and D are linearly-independent parameters dependent on x, y, the Hamiltonian parameters, n_0 and N. The functions F_{ij} are dependent on α, n_0 and N. All are to be published elsewhere. Again, the α-dependent component is sufficient for studying phase transitions. In all that follows, $n_0 = 8$ and $N = 12$, and a cross section of the phase space for these values and for $A = 10$ is shown in Fig. 2 (wherein red dots are points in Fig. 3). Increasing the number of bosons gives phase spaces with the same characteristics. Line types and regions are as in Fig. 1, with the addition of Region 0, in which no minima exist.

Note that this is a more complicated structure than the PACM, and contains an 'end point', beyond which the first order transition ceases to exist. This general structure is maintained at other values of A, making a flat plane of second order phase transition, and a curved plane of first order.

Fig. 3 gives a sample of energy functionals in different regions, the specific points being marked with lower-case letters in Fig. 2. From left to right, the first two panels show graphs either side of a phase transitions, points a) - d). The third shows a graph either side of the dotted line in the Region III of greater B, points e) and f), showing that while each have a barrier, one is bound as $\alpha \to \infty$, and the other is unbound. These tails are dependent on N, with the unbound approaching infinity and the bound negative infinity as $N \to \infty$. Thus, this region is unphysical. The Region III of lesser B, shown in the second panel, has no barrier, and is bound as $\alpha \to \infty$. The fourth panel shows a fully-bound potential with no minima from Region 0, point g).

4. SACM results

Using two spherical clusters simplifies the Hamiltonian. $\alpha+^{16}O$ is such a system, and is examined here. Coefficients for $SU(3)$ were fitted to give the best spectrum in that limit, then those remaining in the $SO(4)$ were fixed in that limit. Then, x was varied with y set to 1. In this way, the best fit for this nucleus is not sought, but phase transitions between symmetry limits are.

The left panel of Fig. 4 shows that a phase transition of second order occurs when $x = 0.6$. The plane is the second order phase transition surface, and the line is the system's path as x varies from 0 to 1, with $y = 1$. The right panel shows the ground state boson expectation, $< n_\pi >$. The kinked line is from a geometrical mapping, and is dependent on the location in α of the deformed minimum. Beyond the phase transition at $x = 0.6$, there is no deformed minimum, and thus the boson expectation goes to the minimum of 8. The smooth line comes from numerical matrix diagonalisation, and has a mixing of states not seen in the geometrical mapping. Note that these results do not go to the cut-off, as was a problem for PACM.

As shown in Fig. 5, the order of low-lying $\alpha+^{16}O$ states in ^{20}Ne is recreated for all values of x,

Journal of Physics: Conference Series **322** (2011) 012010 doi:10.1088/1742-6596/322/1/012010

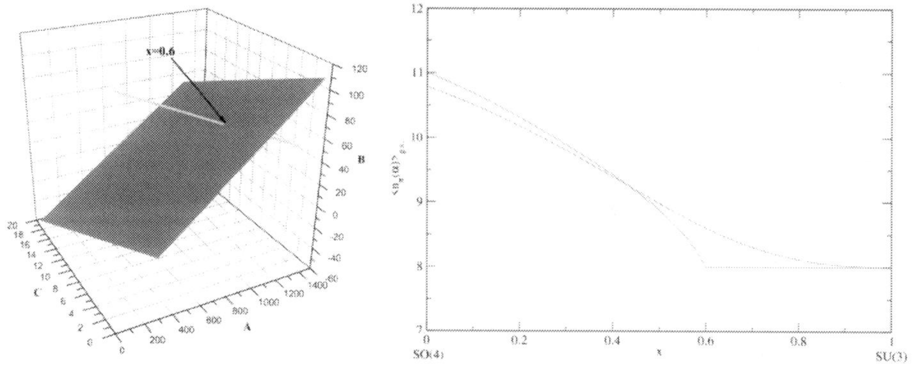

Figure 4. Phase transition at $x = 0.6$ (left), and ground state $< n_\pi >$ (right).

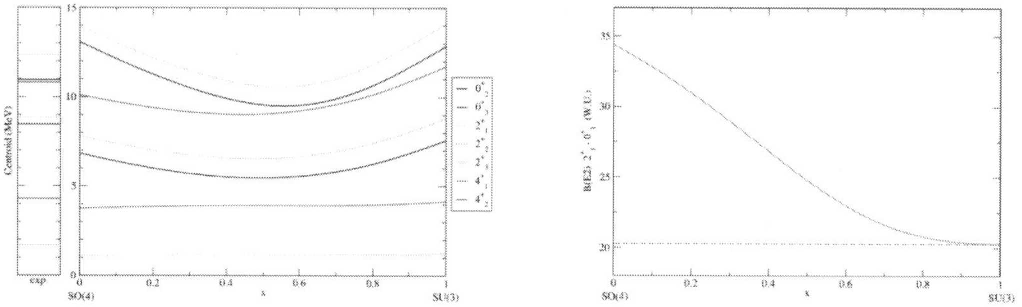

Figure 5. Spectrum of low-lying $\alpha+^{16}O$ states (left panel) and value for $2_1^+ - 0_1^+$ $B(E2)$ transition (right panel) in ^{20}Ne.

and also that the $B(E2)$ value for the $2_1^+ - 0_1^+$ transition is well recreated when compared with experiment (dashed line), especially in the $SU(3)$ realistic for this nucleus. This is a numerical study, and while evidence of the phase transition is seen, it is not strongly so.

5. Conclusions

Differences appearing in the investigation of cluster models which do and do not obey the Pauli exclusion principle (SACM and PACM respectively) have been examined. The Hamiltonian used includes a modified vibron model [3] to describe relative motion, with $\hbar\omega$ fixed by shell model considerations, and the $SU(3)$ model of Elliott [4] for internal cluster structure, rather than the IBA [5, 6]. The Hamiltonian used interpolates between the $SU(3)$ limit, the $SO(4)$ limit, and the $SO(3)$ limit. In order to investigate phase transitions, when moving from one dynamical symmetry limit to another, the coherent state method was used. It was in the coherent states that the Pauli exclusion principle was first neglected and then applied. In the former case, the structure of the coherent state is the simplest.

In both cases, the phase space diagram was constructed in its most general form. Within the PACM only two relevant parameters appear, which in turn are functions of all the parameters of the model. Within the SACM one has three relevant parameters, which are again functions of all the interaction parameters of the theory. Both exhibit first- and second-order phase transitions, but the more complicated SACM phase space includes an 'end line'. Potentials at a selection of illustrative points have been shown for the SACM.

The PACM has been shown to be unrealistic for modelling physical systems, while the SACM has been used successfully to model a physical nuclear system of two spherical clusters, $\alpha+^{16}O$.

Acknowledgments
Support from DGAPA, CONACyT and OTKA (Grant No. K72357) is gratefully acknowledged.

References
[1] Cseh J 1992 *Phys. Lett. B* **281** 173
[2] Cseh J and Lévai G 1994 *Ann. Phys. (N.Y.)* **230** 165
[3] Iachello F 1981 *Phys. Rev. C* **23** 2778
[4] Elliot JP 1958 *Proc. R. Soc. A* **245** 128 and 562
[5] Daley H J and Iachello F 1986 *Ann. Phys. (N.Y.)* **167** 73
[6] Daley H J and Barrett B R 1986 *Nucl. Phys. A* **449** 256
[7] Castaños O, Hess P O, Rochford P and Draayer J P 1991 *Nucl. Phys. A* **524** 469
[8] Hess P O, Lévai G and Cseh J 1996 *Phys. Rev. C* **54** 2345
[9] Cejnar P and Jolie J 2009 *Prog. Part. Nucl. Phys.* **64** 210
[10] Greiner W, Neise L and Stöcker H 1997 *Thermodynamics and Statistical Mechanics* (Heidelberg: Springer)
[11] Cseh J 2010 *J. Phys.: Conf. Ser.* **205** 012021
[12] Cejnar P and Iachello F 2007 *J. Phys. A: Math. Gen.* **40** 581
[13] Frank A and Van Isacker P 1994 *Algebraic Methods in Molecular and Nuclear Structure Physics* (New York: John Wiley and Sons)
[14] Wildermuth K and Tang Y C 1977 *A Unified Theory of the Nucleus* (Braunschweig: Vieweg)

Journal of Physics: Conference Series **322** (2011) 012011 doi:10.1088/1742-6596/322/1/012011

Isovectorial pairing in solvable and algebraic models

Sergio Lerma[1], **Carlos E Vargas**[1] **and Jorge G Hirsch**[2]

[1] Departamento de Física, Facultad de Física e Inteligencia Artificial, Universidad Veracruzana, Lomas del Estadio s/n, C.P. 91000, Xalapa, Veracruz, México

[2] Instituto de Ciencias Nucleares, Universidad Nacional Autónoma de México, AP 70-543, 04510 Mexico DF, Mexico

E-mail: slerma@uv.mx

Abstract. Schematic interactions are useful to gain some insight in the behavior of very complicated systems such as the atomic nuclei. Prototypical examples are, in this context, the pairing interaction and the quadrupole interaction of the Elliot model. In this contribution the interplay between isovectorial pairing, spin-orbit, and quadrupole terms in a harmonic oscillator shell (the so-called pairing-plus-quadrupole model) is studied by algebraic methods. The ability of this model to provide a realistic description of N=Z even-even nuclei in the fp-shell is illustrated with ^{44}Ti. Our calculations which derive from schematic and simple terms confirm earlier conclusions obtained by using realistic interactions: the SU(3) symmetry of the quadrupole term is broken mainly by the spin-orbit term, but the energies depends strongly on pairing.

1. Introduction

Starting from a harmonic mean field and a spin-orbit one-body term, the independent particle shell model of atomic nuclei is able to reproduce the shell closures found experimentally and to describe dominant features of closed shell nuclei. When the number of nucleons deviates from the magic numbers, residual interactions have to be taken into account for a proper description of these nuclei. Two dominant terms of these residual forces are the quadrupole and pairing interactions [1]. A common characteristic of these two-body interactions is that they can be approximated by schematic terms possessing an underlying algebraic structure which can be exploited to simplify their study and to interpret clearer their effects. The pairing plus quadrupole Hamiltonian of Bohr, Mottelson and Belyaev [2, 3] is obtained by bringing together all the previous terms and by supposing a valence space by only one harmonic oscillator shell:

$$\mathbf{H} = C\sum_i \mathbf{l}_i \cdot \mathbf{s}_i - \frac{\chi}{2}\mathbf{Q}\cdot\mathbf{Q} + H_p. \tag{1}$$

The relevance of the different terms in Hamiltonian (1) to describe nuclei in the *sd* and *pf* shells has been the subject of several studies, which make use of realistic interactions to derive their results. In ref. [4] the authors showed that the spin-orbit and quadrupole deformation terms suffice to give a correct description of the wave functions for systems with 4 protons and 4 neutrons in the *pf*, *sdg* and *pfh* shells. Additional terms in the Hamiltonian (pairing terms included) change the spectrum without modifying strongly the wave functions. Similar results were reported in [5] for different nuclei in the lower *pf* shell, there it was found that the spin-orbit splitting is responsible for the breaking of the SU(3) symmetry associated to the quadrupole interaction, and that this breaking has a small influence in the B(E2) transition probabilities, which remain close to the values coming from exact SU(3) calculations. The latter result is a consequence of what has been called quasi SU(3) dynamical symmetry or adiabatic mixing of SU(3) irreps [6, 7]. The role of the different pairing channels was discussed in ref. [8]. In this reference it is shown, in the framework of full pf shell model calculations with a realistic KB3 interaction, that the proton-neutron components of pairing are very relevant to correctly describe the moment of inertia of $N = Z$ nuclear rotors, a conclusion already signaled in ref. [9], where the authors concluded that the failure of the HFB method to describe the moment of inertia of ^{48}Cr can be attributed to an improper treatment of pairing correlations. Even if the previous results are very illustrative and robust, it is desirable to have a framework where they can be understood in simpler terms. The present work goes in that direction by considering a simple Hamiltonian with the minimal ingredients for a realistic description of atomic nuclei.

The interplay between the different terms in Hamiltonian (1) has been studied in refs. [10, 11]. In the first of these, the pairing interaction is shown to induce soft triaxial deformations in, otherwise, SU(3) symmetric nuclei with sharp oblate or prolate deformations. It is worth to mention that the previous works included only a kind of nucleon, consequently only the like particle component of the isovectorial pairing interaction is taken into account. The role of the spin-orbit term in the Elliot model is analyzed in ref. [12]. The authors take into account the isospin degree of freedom (but no pairing terms) to suggest a SU(3)-based truncation procedure to diagonalize Hamiltonians in large spaces.

Two works closely related to the present one can be found in [13, 14]. In the former the isovectorial pairing without proton-neutron component and the spin-orbit term are introduced in the Elliot model. The SU(3) truncation procedure previously mentioned is used to describe the yrast energies of ^{48}Cr. Although the energies obtained in this reference do not reproduce the experimental backbending of ^{48}Cr, the structure of the wave functions in terms of SU(3) irreps presents a change in the yrast states where backbending happens. The authors conclude that the lack of proton-neutron pairing correlations may explain the absence of backbending in the spectrum of the model, a conclusion that is supported by the results given in reference [14], where the role of the different pairing channels is studied by using the same Hamiltonian as eq.(1). In particular it is found that the isovectorial pairing is at the origin of the backbending of ^{48} Cr.

In this contribution all the components of isovectorial pairing are included in addition to the spin-orbit and quadrupole terms in the framework of the Elliot model, i.e., by using a SU(3) basis. The breaking of the SU(3) symmetry by the spin-orbit and pairing terms is studied. The model is employed to describe the yrast states of the (N=Z) ^{44}Ti nucleus. In Section 2 the symmetric limits of the model are briefly discussed, whereas in Section 3 the exact solution of the isovectorial pairing with arbitrary single-particle energies (spin-orbit term in this context) is reviewed. The method used to diagonalize the Hamiltonian in a SU(3) basis is presented in Section 4. Results and some conclusions are given in Sections 5 and 6 respectively.

2. The symmetric limits

A good description of light rotational nuclei in the *sd*-shell is provided by the Elliot model [15]. This model assumes a single shell of the spherical harmonic mean field (η) and a quadrupole-quadrupole residual interaction

$$\mathbf{H}_Q = -\frac{\chi}{2}\mathbf{Q}\cdot\mathbf{Q},$$

which can be expressed as a linear combination of the Casimir operators of the group chain [16] $SU(3) \supset SO(3)$: $\mathbf{Q}\cdot\mathbf{Q} = 4\mathbf{C}_2[SU(3)] - 3\mathbf{C}_2[SO(3)]$, where $\mathbf{Q}_\mu = \sum_l \sqrt{8(2l+1)}[\mathbf{a}^\dagger_{l\sigma\tau}\tilde{\mathbf{a}}_{l\sigma\tau}]^{(200)}_{(\mu 00)}$, with $\mathbf{a}^\dagger_{lm_\sigma\sigma_z\tau\tau_z}$ nucleon creators of angular momentum, spin and isospin l,σ, and τ, $\tilde{\mathbf{a}}$ is the tensorial respective nucleon annihilator, and the sum runs over $l = \eta, \eta - 2, ..., 0$ or 1. In the coupled operator, the superindex (200) refers to the total orbital angular momentum, the total spin and the total isospin, while the subindex (μ00) describes their projections.

On the other hand, the pairing interaction that explains naturally the even-odd staggering of nuclear masses and the gap energy in the spectrum of even-even nuclei, can be modeled by the schematic interaction

$$\mathbf{H}_p = -g_1 \mathbf{b}^\dagger \cdot \tilde{\mathbf{b}} - g_0 \mathbf{D}^\dagger \cdot \tilde{\mathbf{D}},$$

where the dots represent scalar couplings in isospin and spin space respectively. The isovectorial and isoscalar pairs, written in a ls scheme, are

$$\mathbf{b}^\dagger_{t_z} = \sum_l \sqrt{\frac{2l+1}{2}} \left[\mathbf{a}^\dagger_{l\sigma\tau} \times \mathbf{a}^\dagger_{l\sigma\tau}\right]^{(001)}_{(00t_z)}, \qquad \mathbf{D}^\dagger_{s_z} = \sum_l \sqrt{\frac{2l+1}{2}} \left[\mathbf{a}^\dagger_{l\sigma\tau} \times \mathbf{a}^\dagger_{l\sigma\tau}\right]^{(010)}_{(0s_z0)},$$

The isovectorial pairing interaction above, includes the like particles pairing ($\mathbf{b}^\dagger_{t_z}$ with $t_z = -1$ and $t_z = 1$), and the two channels of proton-neutron pairing ($\mathbf{b}^\dagger_{t_z=0}$ and $\mathbf{D}^\dagger_{s_z}$).

It can be shown [17] that the pair creators are the ladder operators of a SO(8) algebra. If $g_1 = g_0$, the pairing Hamiltonian can be expressed in terms of the Casimir operator of that algebra and the model commutes with the generators of the SU(4)-Wigner symmetry. In the case where only isovectorial or isoscalar channels are included, the pairing interaction can be expressed in term of the Casimir operator of a SO(5) algebra, and the model is, respectively, isospin or spin invariant. In all of the previous symmetric cases, the wave function of the ground state for $N = Z$ even-even systems exhibits a quartet structure in the symmetric space. If only one kind of nucleons is allowed, a situation that arise when the valence protons and valence neutrons occupy different shells, the relevant algebra is reduced to the more tractable SU(2) algebra.

The most general case ($g_1 \neq g_0$) has been studied in [18], where it is shown that the quartet structure of the ground state wave function found in the symmetric limits, remains in the following form

$$\left(\cos\theta \mathbf{b}^\dagger \cdot \mathbf{b}^\dagger - \sin\theta \mathbf{D}^\dagger \cdot \mathbf{D}^\dagger\right)^{(M/2)} |0\rangle, \tag{2}$$

where θ is a parameter which takes the values $\theta = 0$ for pure isovectorial pairing, $\theta = \pi/2$ for pure isoscalar pairing and $\theta = \pi/4$ for $g_1 = g_0$.

Before discussing the solution for a Hamiltonian which additionally includes a spin-orbit term, it is convenient to mention the work of ref. [19], where it is shown that the spin-orbit is very effective to suppress the isoscalar correlations, leaving the isovectorial ones unaffected. This result that can be understood by expressing the isovectorial and isoscalar pairs in a jj-scheme, while the isovectorial pairs relates nucleons in the same $j-$ shell, the isoscalar pairs are formed by nucleons in differents j-shells. For that reason, in what follows only the isovectorial pairing will be considered.

3. Exact solution of the isovectorial pairing plus spin-orbit interaction

The isovectorial pairing with the spin-orbit term is a particular case of the isovectorial pairing hamiltonian with arbitrary single-particle energies,

$$\mathbf{H}_{SO(5)} = \sum_i \epsilon_i \left(\mathbf{N_{pi}} + \mathbf{N_{ni}}\right) - g \sum_{ij} \left(\mathbf{b}^\dagger_{1i}\mathbf{b}_{1j} + \mathbf{b}^\dagger_{0i}\mathbf{b}_{0j} + \mathbf{b}^\dagger_{-1i}\mathbf{b}_{-1j}\right), \tag{3}$$

which has been shown to be diagonalizable by a Bethe ansatz in refs.[20, 21]. The energies of the previous Hamiltonian are given by $E = \sum e_s$, where the parameters e_s are the solution of the following set of non-linear equations which involves additionally a set of parameters ω_p :

$$\sum_{q\neq s}^{M} \frac{2}{e_s - e_q} + \sum_{q}^{M_\omega} \frac{1}{\omega_q - e_s} + \sum_{j} \frac{1}{2\epsilon_j - e_s} = \frac{1}{g}, \qquad \sum_{q}^{M} \frac{2}{e_q - \omega_p} + \sum_{q\neq p}^{M_\omega} \frac{2}{\omega_p - \omega_q} = 0,$$

with $s = 1, ..., M$ and $q = 1, ..., M - T$, where M and T are, respectively, the number of pairs and isospin of the eigenstate.

Besides the energy, the parameters e_s determine the orbital and spin structure of the correlated pairs through

$$\mathbf{b}_t^\dagger(e) = \sum_i \frac{\mathbf{b}_{ti}^\dagger}{2\epsilon_i - e},$$

whereas the ω_q parameters determine the wave function structure in isospin space. General expression can be found in [22]. Here we only signal that, for the simplest non-trivial case of $N = Z = 4 = M$, it was shown that the wave function can be approximated by

$$\left(\mathbf{b}^\dagger(e_1) \cdot \mathbf{b}^\dagger(e_2)\right) \left(\mathbf{b}^\dagger(e_3) \cdot \mathbf{b}^\dagger(e_4)\right) |0\rangle,$$

for almost all values of the pairing strength, except for small ranges around the values where one pair changes its orbital character from a normal pair (uncorrelated pairs occupying the single-particle levels) to a Cooper-like pair. The previous expression is reminiscent of the quartet structure found in the case of one completely degenerate shell, eq.(2), but contrary to that case, here the quartets are not identical. Only in the limit $g \to \infty$ (which implies $|e_s| \to \infty$), the effect of the single-particle energies is diluted and the one-shell solution is recovered.

One of the advantages of the Bethe ansatz is that it can be used to explore the solution in very large spaces, where standard diagonalization techniques are unable to manage the problem. As an illustration of this, in figure 1 a graphical representation of the solution is shown for the case of 44 nucleons ($N = Z = 22$) moving in a set of single particle energies given by the first six harmonic oscillator shells ($\eta = 0, \ldots, 5$) and a spin-orbit term. The figure shows two cases, the first one, corresponding to weak paring strength where all the pairs are associated to the lowest single-particle levels, represents the case of uncorrelated pairs formed by nucleons occupying the single-particle levels. In the second case, only the deepest pairs remain associated to the single particle levels, the pairs in the upper part expand in the complex plain, signaling the formation of a superconducting state whose structure in isospin space is very difficult to determine.

4. Isovectorial pairing interaction in a SU(3) basis

The quadrupole term breaks the exact solvability of the pairing plus spin-orbit Hamiltonian, therefore a different procedure is needed. The results of refs.[4, 5] indicates the most convenient way to proceed: rather than including the quadrupole as a perturbation in the previous exact solution, it is better to include the previous Hamiltonian in a basis which diagonalizes the quadrupole-quadrupole term. This basis is given by the following chain of groups [12, 23]:

$$\begin{array}{ccccc} [1^{N_\tau}] & & [\tilde{f}_\tau] \quad [f_\tau] & \alpha_\tau & (\lambda_\tau, \mu_\tau) s_\tau \\ U(\Omega_\tau) & \supset & U\left(\frac{\Omega_\tau}{2}\right) \times U(2) & \supset & SU(3) \times SU(2), \end{array}$$

where $\tau = p, n$ and Ω_τ is the degeneracy of the harmonic oscillator shell. The previous classification applies to protons and neutrons separately, the basis we use is obtained by coupling the proton and neutron SU(3) irreps to a total SU(3) irrep:

$$|\{N_p[f_p]\alpha_p(\lambda_p, \mu_p)s_p\}\{N_n[f_n]\alpha_n(\lambda_n, \mu_n)s_n\}\rho(\lambda, \mu)\kappa LS; JM\rangle,$$

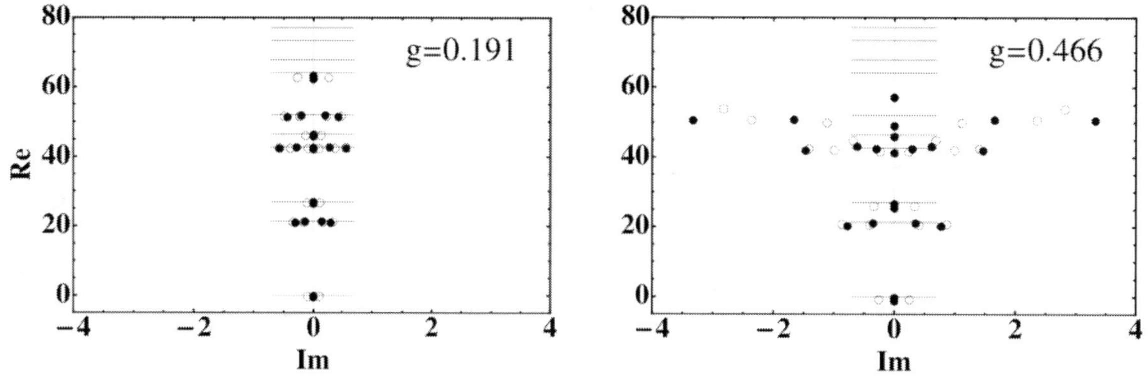

Figure 1. Graphical representation of the solution in the complex plain for two cases: weak (left) and strong (right) isovectorial pairing. Black and white circles represent respectively e and ω parameters. Horizontal lines are twice the single-particle energies of the spherical shell model.

where ρ and κ are multiplicity labels which count respectively, the number of times the irrep (λ, μ) occurs in the product $(\lambda_p, \mu_p) \otimes (\lambda_n, \mu_n)$ and the number of occurrences of the angular momentum L in the irrep (λ, μ).

To calculate the matrix elements of the Hamiltonian in the previous basis, we need to express the pairing and spin-orbit terms as a linear combination of SU(3) tensors. With these expressions and eq. (10) of ref.[23], which derives directly from the Wigner-Eckart theorem, the matrix elements are expressed in terms of reduced matrix elements, which, in turn, can be calculated employing the code described in ref.[24]. The expression for the spin-orbit and like particles pairing can be found in refs.[12, 23], the only term that remains to be calculated is the proton-neutron component of isovectorial pairing. The result is given here

$$\begin{aligned} H_{PN} &= g\mathbf{b}^{\dagger}_{t_z=0}\tilde{\mathbf{b}}_{t_z=0} = -\frac{g}{2}\sum_{\lambda_p\lambda_n}\sum_{\lambda_o\mu_o\rho_o}\sum_{s=0}^{1}\sqrt{2s+1}C(\lambda_p,\lambda_n,\lambda_o,\mu_o,\rho_o) \\ &\left[\left[\mathbf{a}_{p(\eta,0)1/2}\tilde{\mathbf{a}}_{p(0,\eta)\frac{1}{2}}\right]^{(\lambda_p,\lambda_p)s}\left[\mathbf{a}_{n(\eta,0)1/2}\tilde{\mathbf{a}}_{n(0,\eta)\frac{1}{2}}\right]^{(\lambda_p,\lambda_p)s}\right]^{\rho_o(\lambda_o,\mu_o)\kappa_o=1L_o=0,S_o=0}, \end{aligned} \tag{4}$$

where the function C is given by SU(3) Clebsch-Gordan coefficients

$$\begin{aligned} C(\lambda_p,\lambda_n,\lambda_o,\mu_o,\rho_o) = \sum_{ll'}\sum_{L\kappa_p\kappa_n}\sqrt{2L+1}\langle(\eta,0)l(0,\eta)l'||(\lambda_p,\lambda_p)\kappa_pL\rangle \\ \langle(\eta,0)l(0,\eta)l'||(\lambda_n,\lambda_n)\kappa_nL\rangle \\ \langle(\lambda_p,\lambda_p)\kappa_pL(\lambda_n,\lambda_n)\kappa_nL||(\lambda_o,\mu_o)\kappa=1L=0\rangle_{\rho_o}. \end{aligned} \tag{5}$$

5. Results for ^{44}Ti

The strength ratios of the different forces in the Hamiltonian are fixed by systematics [1, 25]:

$$\frac{C}{\epsilon_o} = 2\hbar\omega_o\kappa \approx \frac{(0.16)41}{A^{1/3}}[MeV], \quad \frac{\chi}{\epsilon_o} = 28/A^{5/3}[MeV], \quad \frac{g}{\epsilon_o} = 19.51/A[MeV].$$

where the parameter ϵ_o is a dimensionless, global parameter that only fix the energy scale without modifying the wave functions.

Journal of Physics: Conference Series **322** (2011) 012011 doi:10.1088/1742-6596/322/1/012011

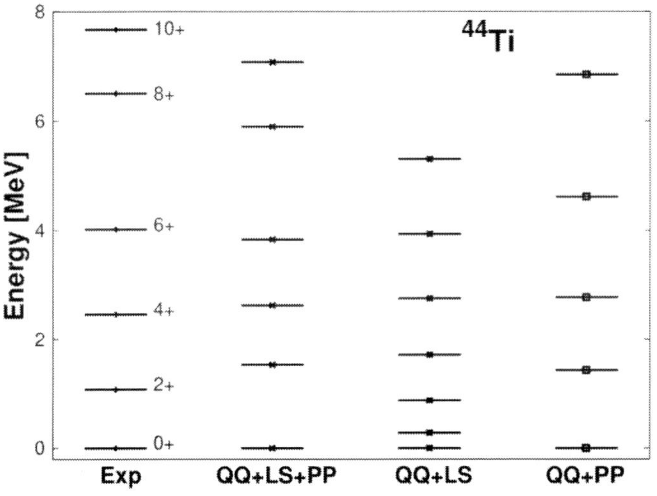

Figure 2. Yrast spectrum of ^{44}Ti, the first column shows the experimental values, the last three columns are obtained from the model Hamiltonian with all (2nd column), quadrupole+spin-orbit (3rd column), and quadrupole+pairing (4th column) terms.

We use a system corresponding to ^{44}Ti to illustrate our results. This system contains two valence protons and two valence neutrons moving in the $\eta = 3$ oscillator shell, where the spin orbit-term produces 4 single particle-levels ($f_{7/2}, p_{3/2}, p_{1/2}, f_{5/2}$). The dimension of the space allows a full diagonalization, however a SU(3) dictated truncation can employed in nuclei with larger spaces. To probe the procedure and the written code, we compare the energies obtained for a Hamiltonian without quadrupole term with the energies coming from the exact solution for $\epsilon_o = 1$. Results are shown in table 1 for the ground state energy (measured respect to the zero energy of the oscillator) and for the excitation energy of the first seniority two state ($J = 2, 4, 6$). The $SU(3)$ code reproduces perfectly the energies and degeneracies obtained from the exact solution.

In the first and second columns of figure 2, the spectrum of the model is compared with the experimental energies of the yrast states of ^{44}Ti. The overall parameter ϵ_o was fixed at 0.7. In the third and fourth columns of the same figure, the spectra of the Hamiltonians without pairing and without spin-orbit term are shown. Note that the quadrupole plus spin-orbit Hamiltonian exhibits a rotational spectrum and that the non-rotational character of ^{44}Ti spectrum is reproduced by the pairing term. From these results one could naively conclude that pairing is the responsible of the SU(3) symmetry breaking of the quadrupole term, a wrong conclusion which is refuted by the evidences coming from analyzing the respective wave functions. In figure 3 the contribution of the different SU(3) irreps to the ground state wave functions is shown for the three cases of figure 2. From this figure, it is clear that the SU(3) fragmentation is mainly dictated by the spin-orbit term, whereas the pairing term produces a wave function

	SU(3) Code	Exact Solution
GS energy ($J = 0$)	192.54 MeV	192.54 MeV
ΔE_2 ($J = 2, 4, 6$)	2.3905 MeV	2.3904 MeV

Table 1. Comparison between SU(3) code and exact solution results for the ground-state energy and excitation energy of the first seniority two state, for an isovectorial pairng plus spin-orbit Hamiltonian.

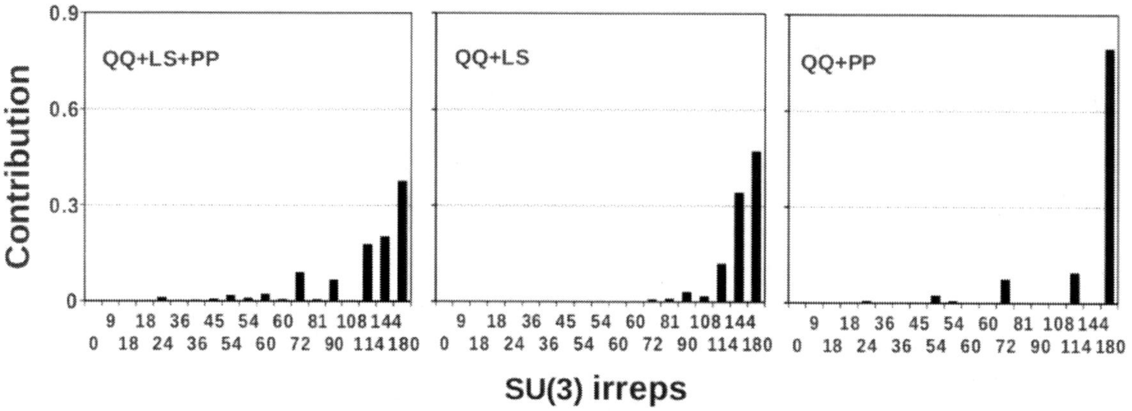

Figure 3. Contribution of the different SU(3) irreps to the ground-sate wave functions for the same Hamiltonian as in figure 2. In the horizontal axis the different SU(3) irreps are represented by their Casimir values

dominated completely (80 %) by the SU(3) irrep with largest SU(3) casimir value (which is the ground-state wave function in the pure quadrupole SU(3) symmetric limit). Figures 2 and 3 illustrates the way as the SU(3) symmetry is broken in the case of realistic values of the Hamiltonian parameters: the spin-orbit term is the responsible of the SU(3) breaking even if its influence in the spectrum is rather marginal, on the contrary the pairing term breaks slightly the SU(3) symmetry of the quadrupole term, but the spectrum depends strongly on it. The SU(3) fragmentations of the ground-state wave functions shown in figure 3 are very similar to those obtained in [5] for a realistic Kuo-Brown-3 realistic interaction, which indicates the ability of the model to properly describe real systems despite its simplicity. As the total angular momentum increases, the effect of pairing is reduced due to the Coriolis force which breaks the Cooper pairs. It was verified that the effect of pairing disappears for $J \geq 8$ (seniority four states), a result that can be understood from the fact that these states are obtained by breaking all the pairs of the $f_{7/2}$ subshell, the lowest in energy.

6. Conclusions

Some aspects of isovectorial pairing interaction in the context of an exactly solvable (pairing+spin-orbit) and an algebraic (pairing+quadrupole model) model were discussed. The isovectorial pairing plus spin-orbit term is solvable by a Bethe ansatz, which allows to explore very large systems untractable by other standard methods. The graphical representation of the solution allows a qualitative description of the wave function, but quantitative results are much more challenging. For the simplest non-trivial case ($N = Z = 4$), it was shown that the quartet structure in isospin space found in the model with only one degenerate level, survives the splitting of the single-particle levels for almost all values of the pairing strength. For the Hamiltonian which includes additionally the quadrupole interaction, it was found that, for realistic values of the Hamiltonian parameters, the pairing interaction acts like a "quasi-scalar" tensor, that is to say it breaks slightly the SU(3) symmetry but has a strong effect in the energy spectrum. On the contrary the spin-orbit term determines the way as the SU(3) symmetry of the quadrupole term is broken, but leaves the rotational spectrum of the quadrupole term almost unmodified. The present study can be extended to other $N = Z$ nuclei in the pf shell, in particular to reproduce the experimental backbending of ^{48}Cr. Work in this direction is underway.

Acknowledgments

SLH was supported by the Mexican Federal Program for the Improvement of Professors under Project No F-PROMEP-38/Rev-03, SEP-23-005. This work was supported in part by Conacyt, México, by FONCICYT project 94142, and by DGAPA, UNAM.

References

[1] Dufour M and Zuker A P 1996 *Phys. Rev.* C **54** 1641
[2] Bohr A, Mottelson B R and Pines D 1958 *Phys. Rev.* **110** 936
[3] Belyaev S T 1959 *Mat. Fys. Medd. Dan. Vid. Selsk* **31, No. 11** 936
[4] Zuker A P, Retamosa J, Poves A and Caurier E 1998 *Phys. Rev. C* **52** 485
[5] Gueorguiev V G, Draayer J P and Johnson C W 2001 *Phys. Rev. C* **63** 014318
[6] Rochford P and Rowe D J 1998 *Phys. Lett. B* **210** 5
[7] Bahri C, Rowe D J and Wijesundera W 1998 *Phys. Rev. C* **58** 31539
[8] Poves A and Martinez-Pinedo G 1998 *Phys. Lett. B* **430** 203
[9] Caurier E, Egido J L, Martinez-Pinedo G, Poves A, Retamosa J, Robledo L M and Zuker A P 1995 *Phys. Rev. Lett.* **75** 2466
[10] Bahri C, Escher J and Draayer J P 1995 *Nucl. Phys. A* **592** 171
[11] Escher J, Bahri C, Troltenier D and Draayer J P 1998 *Nucl. Phys. A* **633** 662
[12] Vargas C, Hirsch J G, Hess P O and Draayer J P 1998 *Phys. Rev. C* **58** 1488
[13] Hirsch J G, Hess P O, Hernández L, Vargas C, Beuschel T and Draayer J P 1999 *Rev. Mex. Fís.* **45 S2** 86
[14] Yang L, Zhao Y M, Pittel S, Thakur B, Sandulescu N and Poves A 2010 *Journal of Physics: Conference Series* **239** 012008
[15] Elliot J P 1958 *Proc. R. Soc. Lond.* A **245** 128; 562
[16] Frank A, Jolie J and Isacker P V 2009 *Symmetries in atomic nuclei* (New York: Springer)
[17] Flowers B H and Szpikowski S 1964 *Proceedings of the Physical Society* **84** 673
[18] Palchikov Y V, Dobes J and Jolos R V 2001 *Phys. Rev. C* **63** 034320
[19] Juillet O and Josse S 2000 *Eur. Phys. J. A* **8** 291
[20] Links J, Zhou H Q, Gould M D and McKenzie R H 2002 *J. Phys. A* **35** 6459
[21] Dukelsky J, Gueorguiev V G, van Isacker P, Dimitrova S, Errea B and Lerma-H S 2006 *Phys. Rev. Lett* **96** 072503
[22] Lerma-H S 2010 *Phys. Rev. C* **82** 014302
[23] Troltenier D, Bahri C and Draayer J P 1995 *Nucl. Phys. A* **586** 53
[24] Bahri C and Draayer J P 1994 *Comp. Phys. Comm* **83** 59
[25] Ring P and Schuck P 1980 *The nuclear many-body problem* (New York: Springer-Verlag)

Journal of Physics: Conference Series **322** (2011) 012012 doi:10.1088/1742-6596/322/1/012012

Quantum phase transitions in finite algebraic systems

Pavel Cejnar
Institute of Particle and Nuclear Physics, Faculty of Mathematics and Physics, Charles University, V Holešovičkách 2, 180000 Prague, Czech Republic
E-mail: cejnar@ipnp.troja.mff.cuni.cz

Abstract. We discuss features of quantum phase transitions affecting collective dynamics of many-body systems. Due to a finite number of collective degrees of freedom, the infinite-size limit of the corresponding models coincides with the classical limit and quantum critical phenomena for both the ground and excited states are determined by the structure of the classical phase space. An example of an excited-state phase transition for the interacting boson model in its SU(3) limit is presented.

1. Introduction

Quantum phase transitions (QPTs) belong to the most interesting topics in modern quantum physics [1]. They were recognized in the 1970's [2, 3] as abrupt changes of the ground state of some condensed-matter systems with a varying non-thermal control parameter λ (typically an interaction strength in the Hamiltonian). The ground-state wave functions on the two sides of the quantum critical point λ_{c0} constitute two qualitatively different types of solution which—for the system's size increased to infinity—cannot be smoothly connected. This shows up also in a non-analytic change of the ground-state energy at the critical point.

Various types of QPT have been theoretically described and in numerous cases also experimentally realized in diverse solid-state and optical systems [1, 4, 5]. QPTs are also discussed in connection with the equilibrium shapes of atomic nuclei [3, 6] (for recent reviews see Refs. [7, 8]) and with the form of molecules and various mesoscopic systems [9]. Although nuclei and molecules as strictly finite objects can only show some faint precursors of the ideal QPT behavior, they play an important role in the development of the concept. In particular, since the models describing these systems are often formulated in an algebraic way [10, 11, 12, 13], they bring to the field descriptions based on exact symmetries and their various effective approximations [14].

In this article, we focus on QPTs in the systems with finite (moderate) numbers of degrees of freedom. In many-body physics, such finite systems (or, more precisely, finite models) are mostly used for the description of *collective dynamics*. The number of collective degrees of freedom of a given type (e.g., those describing quadrupole deformation of nuclei) is fixed and does not depend on the number of constituents of the many-body system (the number of nucleons). As argued below, such models and their QPTs have some specific features.

2. Finite versus infinite systems

Let us consider an algebraically based quantum model for some collective degrees of freedom. The Hilbert space of state vectors coincides with one of the irreps of a finite dynamical (spectrum generating) algebra $\mathcal{G}$. The generators of $\mathcal{G}$, satisfying the closure relation $[G_i, G_j] = c_{ijk}G_k$ (where c_{ijk} stand for structure constants), serve as building blocks for all relevant physical operators, including the Hamiltonian $H = H(G_i)$.

Assume that the model includes a parameter $\aleph$ measuring the size of the system. For instance, in the interacting boson model of nuclei as well as in the related vibron models of molecules [10, 11, 12] the size parameter can be naturally identified with the total number of bosons $N_{\rm B}$. In general, it is convenient if the thermal dispersion $\langle\langle E^2\rangle\rangle \equiv \langle E^2\rangle - \langle E\rangle^2$ of the total energy grows linearly with $\aleph$ (note that $\langle\langle E^2\rangle\rangle$ is proportional the total specific heat of the system). Then thermal fluctuations of the scaled energy $\mathcal{E} \equiv E/\aleph$ vanish for $\aleph \to \infty$, allowing one to properly define the thermodynamic limit with a crossover from canonical to microcanonical description.

Due to the finiteness of the dynamical group $\mathcal{G}$, the system has a finite number of degrees of freedom f. This number coincides with the size of a complete set of not fully degenerate commuting operators characterizing $\mathcal{G}$ [15]. The fact that f does not change with $\aleph$ implies an important difference from ordinary many-body systems, in which the number of degrees of freedom grows proportionally to the number of constituents. In particular, for finite algebraic models the *infinite-size* limit is synonymous with the *classical* limit. This can be deduced from the use of scaled generators $g_i = G_i/\aleph^\kappa$ with $\kappa > 0$ defined through a consistent scaling of the Hamiltonian $H/\aleph = H(G_i/\aleph^\kappa)$. The closure relation for G_i leads for $\aleph \to \infty$ to $[g_i, g_j] \to 0$. Since vanishing commutators of scaled observables constitute a classical description, we are here in a radically different situation than with a truly many-body model, for which the infinite-size limit leads to a quantum-field description.

3. Ground-state and excited-state phase transitions

The classical limit of finite collective models is standardly obtained by the coherent-state formalism [16]. There is a plethora of ways how the coherent states can be constructed for the dynamical algebra $\mathcal{G}$, the resulting classical Hamiltonians being equivalent up to some canonical transformations. One possibility is to employ a bosonic representation of $\mathcal{G}$ (Holstein-Primakoff or Schwinger type) and then use either Glauber coherent states with indefinite total boson number $N_{\rm B}$, or their projection to a fixed $N_{\rm B}$ (boson condensate states). It turns out that every independent boson operator $b_k^\dagger$ that appears in the representation adds one degree of freedom, the resulting number being reduced by one due to the conservation of $N_{\rm B}$. The corresponding coordinate q_k and momentum p_k are associated with the complex coefficient in front of $b_k^\dagger$ in the respective coherent-state expression. Denoting the coherent state as $|\{p_k, q_k\}\rangle \equiv |p, q\rangle$, the classical Hamiltonian is given by

$$H_{\rm cl}(\lambda; p, q) = \aleph^{-1}\langle p, q|H(\lambda)|p, q\rangle, \tag{1}$$

where the scaling ensures convenient properties of $H_{\rm cl}$ in the asymptotic domain of $\aleph$.

Any critical behavior of the ground-state energy can only be observed for $\aleph \to \infty$. It results from a non-analytic dependence of the energy minimum $\mathcal{E}_0(\lambda) = \min_{p,q} H_{\rm cl}(\lambda; p, q)$ on the control parameter λ at a certain critical value $\lambda = \lambda_{c0}$. Note that the minimization of $H_{\rm cl}$ in p and q can be reduced to minimization of the classical potential $V_{\rm cl}(\lambda; q) \equiv H_{\rm cl}(\lambda; p = 0, q)$ in q. Within the coherent-state formalism, $V_{\rm cl}$ can be expressed as a power series in both q and λ. This makes it possible, in finite models, to describe the ground-state QPTs analytically, using either the Landau theory of thermal phase transitions [17] or the catastrophe theory of structurally unstable systems [4].

Non-analytic dependences can also affect individual excited states [18, 19, 20]. The critical value λ_{ci} associated with a given energy level $\mathcal{E}_i(\lambda) = E_i(\lambda)/\aleph$ depends on the excitation index $i \geq 0$ counting the states increasingly with energy. It should be stresses that these so-called excited-state quantum phase transitions (ESQPTs) do not affect only the states in a close vicinity

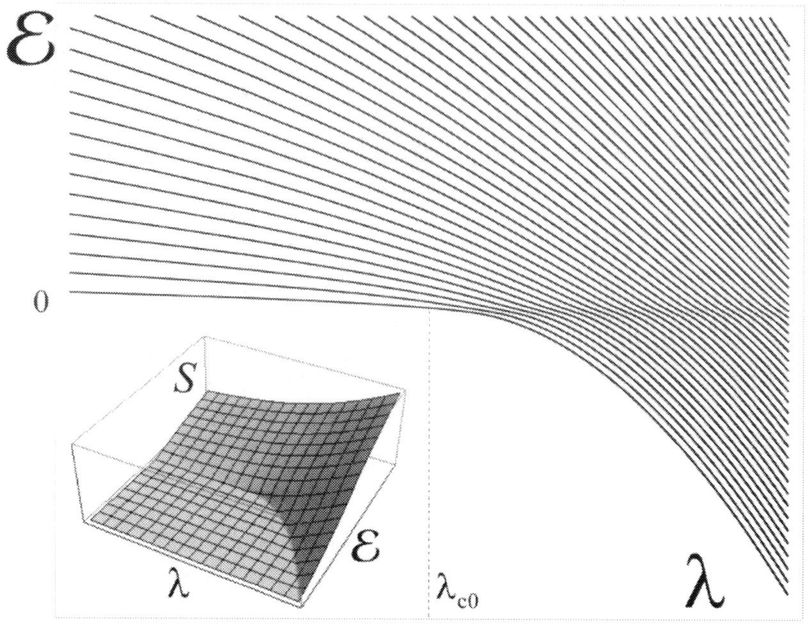

Figure 1. Contour lines (a surface plot in the inset) of the action S per periodic orbit for the classical Hamiltonian (2). The units of all axes are arbitrary (scaling dependent). The point λ_{c0} corresponds to the ground-state QPT while the cumulation of contours at $\mathcal{E} = 0$ for $\lambda > \lambda_{c0}$ indicates a chain of ESQPTs.

of the ground state $i = 0$, but can be observed across the whole spectrum of quantum states or its considerable fraction. In finite models, the ESQPTs are rooted in non-analyticities connected with *dynamical* features of the classical Hamiltonian $H_{\rm cl}$. This is in contrast to the ground-state QPTs, which reflect properties of the *static* equilibrium.

The ESQPTs were so far mostly studied [18, 19, 20] in systems with just one effective degree of freedom ($f = 1$). Such systems provide a straightforward algorithm to analyze properties of the scaled energies $\mathcal{E}_i(\lambda)$ as functions of the control parameter λ for $\aleph \to \infty$. It follows from the familiar semiclassical quantization formula $S(\lambda, \mathcal{E}_i) = {\rm const}_i$, where $S = \oint p\,dq$ represents the action along a closed orbit. The constant on the right-hand side is the $(i + m/4)$th multiple of the Planck constant h, where the integer m (called Maslov index) depends on the particular orbit type.

Consider as an example a Hamiltonian

$$H_{\rm cl} = \frac{1}{2M}\,p^2 + (\lambda_{c0} - \lambda)\,q^2 + q^4 \qquad (2)$$

where M is an arbitrary mass and λ_{c0} a critical value of the control parameter λ. Hamiltonians of this form (with a quartic potential, but usually with a more complicated kinetic terms) appear in the classical limit of some finite bosonic systems with two-body interactions. The contour curves S=const in the plane $\lambda \times \mathcal{E}$ are shown in Fig. 1 for $\lambda \in [0, 2\lambda_{c0}]$. According to the semiclassical quantization, the contours of S represent spectral "flow lines" moving in parallel with actual energy levels in the limit $\aleph \to \infty$. Note that the Hamiltonian (2) implies a ground-state QPT in the corresponding quantum system at $\lambda = \lambda_{c0}$, where the second derivative of $\mathcal{E}_0(\lambda)$ changes discontinuously. This can be anticipated from the trajectory of the lowest contour in Fig. 1. For λ increasing above λ_{c0}, a chain of ESQPTs is observed for excited states with consecutively

Journal of Physics: Conference Series **322** (2011) 012012 doi:10.1088/1742-6596/322/1/012012

increasing values of $i = 1, 2, 3, \ldots$. This can be seen from the coherent bending of contours around $\mathcal{E} = 0$. The semiclassical analysis shows [18] that the second derivative of each $\mathcal{E}_i(\lambda)$ diverges logarithmically (while the first derivative drops to zero) at the point λ_{ci} where the level crosses a critical energy $\mathcal{E}_c = 0$. This energy corresponds to the local maximum ($q = 0$) of the double-well potential in Eq. (2) for $\lambda > \lambda_{c0}$. The classical trajectory in the phase space at $\mathcal{E} = \mathcal{E}_c$ has an infinite period and forms a *separatrix* between two different types of orbits—"bounded" (i.e., located within one of the minima) for $\mathcal{E} < \mathcal{E}_c$, and "unbounded" (flying over both minima) for $\mathcal{E} > \mathcal{E}_c$.

For non-separable systems with $f > 1$ the determination of the $\mathcal{E}_i(\lambda)$ curves by a simple semiclassical formula fails. In fact, if an ESQPT affects a bunch of excited states in a non-integrable system, only an ensemble-averaged dependence $\langle\mathcal{E}_i(\lambda)\rangle$ is relevant since individual levels often show rather different responses [20]. In these cases, the phase-space volume criterion becomes more useful. It is based on the integral

$$\Gamma(\lambda, \mathcal{E}) = \int \delta(\mathcal{E} - H_{\rm cl})\, d^f q\, d^f p \tag{3}$$

with $\Gamma\, d\mathcal{E}$ standing for the phase-space volume available in a $d\mathcal{E}$ interval around energy $\mathcal{E}$. A non-analytic dependence of Γ on $\mathcal{E}$ at a certain value $\mathcal{E}_c(\lambda)$ for a fixed λ indicates an ESQPT at this critical energy. Note that Γ is proportional to the smooth part $\bar{\rho}$ of the total level density $\rho(\lambda, E)$ (the oscillatory part $\tilde{\rho}$ yielding infinitely rapid fluctuations in the $\aleph \to \infty$ limit). Since $\ln \rho$ represents the microcanonical entropy, the non-analyticities of the phase-space volume (3) can be translated into the language analogous to that used in the description of thermal phase transitions. It should be stressed, however, that systems with moderate phase-space dimensions can generate phase transitions of different types than those known for $f \to \infty$ systems. For instance, a local maximum of a one-dimensional potential, such as the one in Eq. (2), yields an infinite peak in Γ and ρ (cf. the $\mathcal{E} = 0$ bunching of contours in Fig. 1), which from the thermodynamic viewpoint is a rather nongeneric type of dependence.

4. Example: IBM in the SU(3) limit

Ground-state QPTs were thoroughly investigated in several models describing collective dynamics—see Refs. [7, 8] for recent reviews focused mostly on nuclear physics. On the other hand, the excited-state QPTs have been so far described only in the simplest situations. The existing studies refer mostly to systems with an effective number of degrees of freedom $f = 1$, like the Lipkin model [21], the two- and three-dimensional vibron models for a subset of states with angular momentum $l = 0$ [22], the interacting boson model along its O(6)-U(5) transition for a subset of states with angular momentum and seniority $l = v = 0$ [18], a so-called cusp Hamiltonian [20], or schematic pairing models [19, 23, 24]. New results concern the Jaynes-Cummings model of quantum optics and a related model based on the SU(1,1) dynamical algebra [25]. In majority of cases, a singularity of the type shown in Fig. 1 was reported. It is connected with a local maximum (or a saddle point) of one-dimensional classical potential.

Here we present another example of an ESQPT, which we derive for the SU(3) limit of the interacting boson model (IBM) [10]. The Hamiltonian is taken in the simple form

$$H = -N_{\rm B}^{-1}(Q \cdot Q), \quad Q = s^\dagger d + d^\dagger s - \frac{\sqrt{7}}{2}[d^\dagger d]^{(2)}, \tag{4}$$

where Q is the SU(3) quadrupole operator. The energy eigenvalues are given in terms of the SU(3) quantum numbers (λ, μ) [10]. The scaled energy $\mathcal{E} = E/N_{\rm B}$ (an additional scaling to the one in H) satisfies $-2 \lesssim \mathcal{E} \lesssim 0$ and is an intensive quantity with respect to $N_{\rm B}$. In the following,

Journal of Physics: Conference Series **322** (2011) 012012 doi:10.1088/1742-6596/322/1/012012

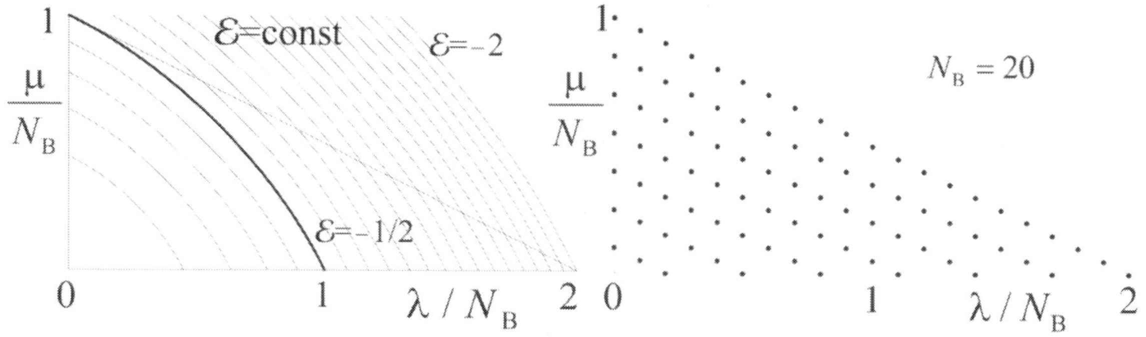

Figure 2. The $\lambda \times \mu$ plane of the SU(3) quantum numbers. Right: Allowed (λ, μ) combinations for the IBM states with $l = 0$ and $N_{\rm B} = 20$. Left: Contours of constant energy $\mathcal{E}$ (step $\Delta\mathcal{E} = 0.1$) from Eq. (5) for $N_{\rm B} \to \infty$. The thick curve represents the critical ESQPT contour $\mathcal{E}_c = -0.5$.

we only consider states with angular momentum $l = 0$. For these states, the formula for $\mathcal{E}$ reads as

$$\mathcal{E} = -\frac{1}{2}\left[\left(\frac{\lambda}{N_{\rm B}}\right)^2 + \left(\frac{\mu}{N_{\rm B}}\right)^2 + \frac{\lambda}{N_{\rm B}}\frac{\mu}{N_{\rm B}} + \frac{3}{N_{\rm B}}\left(\frac{\lambda}{N_{\rm B}} + \frac{\mu}{N_{\rm B}}\right)\right] \tag{5}$$

where the allowed values of $\lambda \in [0, 2N_{\rm B}]$ and $\mu \in [0, N_{\rm B}]$ are given by the known selection rules [10]. These values form a triangle in the $\lambda \times \mu$ plane, which is for $N_{\rm B} = 20$ shown in the right-hand panel of Fig. 2. The contours of constant $\mathcal{E}$ in this plane are for $N_{\rm B} \to \infty$ depicted in the left-hand panel.

Each of the (λ, μ) points in Fig. 2 (right) represents a single eigenstate with $l = 0$. The ground state corresponds to the rightmost point, while the states with an increasing excitation energy fill the interior areas of the triangle towards the vertical side on the left. Note that the points in the triangle have a uniform density $\propto N_{\rm B}^2$. It means that the number of states between energies $\mathcal{E}_1$ and $\mathcal{E}_2$ is proportional to the area of the segment of the triangle located between the respective contours of Eq. (5). The proportionality is exact for $N_{\rm B} \to \infty$. Therefore, the energy dependence of the $l = 0$ state density for large $N_{\rm B}$ can be calculated analytically as $\rho \propto dA/d\mathcal{E}$, where $A(\mathcal{E})$ is the area within the triangle located on the right-hand side of the contour with energy $\mathcal{E}$.

This function is depicted in the left-hand panel of Fig. 3. One observes a nonanalyticity (a place of the vertical-to-horizontal turnover of the tangent), which corresponds to a critical contour with $\mathcal{E} = -0.5$ crossing the upper corner of the triangle (the thick curve in Fig. 2; the infinite derivative of ρ is due to a locally parallel direction of the critical contour with the upper edge of the triangle). The right-hand panel of Fig. 3 shows a finite $N_{\rm B}$-realization of this dependence. The curve, which oscillates around the average predicted in the left-hand panel, represents a smoothened state density obtained from the actual $l = 0$ spectrum for $N_{\rm B} = 60$ (shown on the rightmost edge of the figure).

The classical potential assigned to the above-given SU(3) Hamiltonian by the coherent-state technique is

$$V = -\left(4\beta^2 + 2\sqrt{2}\beta^3 \cos 3\gamma + \frac{1}{2}\beta^4\right)\left(1 + \beta^2\right)^{-2}, \tag{6}$$

where $\beta \in [0, \infty)$ and $\gamma \in [0, 2\pi)$ are the quadrupole shape variables present in the coherent-state parametrization [10]. These variables are the only active degrees of freedom associated with the IBM for $l = 0$ since in the non-rotational case the three Euler angles are frozen. The potential takes values between $V_{\rm min} = -2$ and $V_{\rm max} = 0$ coinciding with the boundaries of the scaled

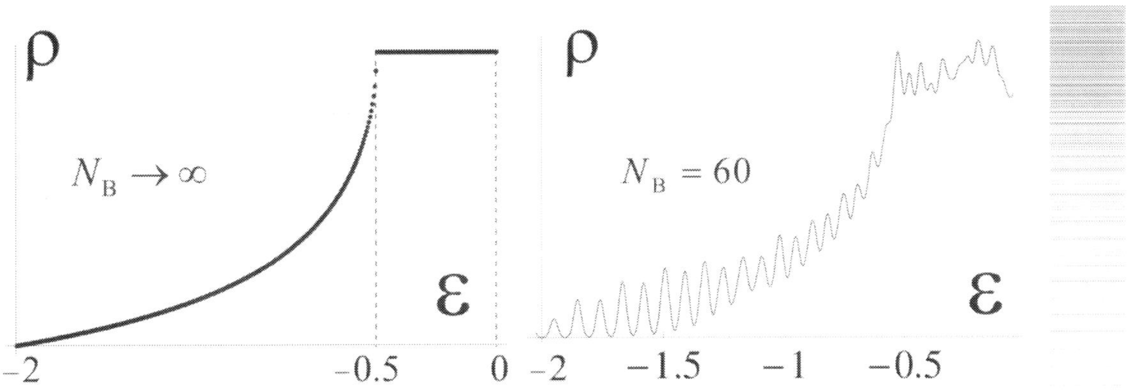

Figure 3. The density of $l = 0$ states in the SU(3) limit of the IBM. Left: The $N_{\mathrm{B}} \to \infty$ density calculated from the area between contours in Fig. 2. Right: Numerical (smoothened) density for $N_{\mathrm{B}} = 60$ (the discrete spectrum shown on the right). Both axes in arbitrary units.

spectrum for $\aleph \to \infty$. The critical value $\mathcal{E}_c = -0.5$ coincides with the saddle point $V_{\mathrm{sad}} = -0.5$ of the potential, demarcating the energy where "unbounded" trajectories ranging over all three potential wells (around minima at $\gamma = 0$, $2\pi/3$, and $4\pi/3$) emerge in the classical dynamics. (Note that the same energy is also associated with the $\beta \to \infty$ asymptote of V.) Therefore, like in the above-discussed case of Hamiltonian (2), a stationary point of the IBM potential (6) is again connected with an ESQPT phenomenon. The analysis of the phase-space volume (3) confirms that a saddle point of a general two-dimensional potential results in a singular growth of $\Gamma(\mathcal{E})$ with an infinite derivative at the corresponding energy $\mathcal{E}_c$. This is clearly manifested in the quantum spectrum of Fig. 3.

Similar considerations can be extended to higher-dimensional models, where however the ESQPT signatures (types of non-analyticities) get softer with an increasing dimension. An interesting question is also connected with the influence of chaotic spectral correlations away from the dynamical symmetries. For example, the IBM coherent-state potential exhibits local maxima and saddle points within the whole Casten triangle, but the associated ESQPTs are less apparent than in the dynamical-symmetry cases (at least in the available spectra with moderate N_{B} values).

5. Conclusions

We have discussed quantum phase transitions in some simple models related to collective dynamics of many-body systems. Specific features of critical phenomena in these models follow from a rather limited number of degrees of freedom f, which does not grow with the size parameter $\aleph$ of the system. As a consequence, all non-analytic dependences of the ground state and excited states on the control parameter λ for $\aleph \to \infty$ originate in some singular properties of the corresponding classical Hamiltonians.

Apart from the ground-state phase transitions, connected with non-analytic changes of the classical point of static equilibrium, we have seen some examples of excited-state phase transitions, which are rooted in abrupt changes of classical dynamics. For atomic nuclei, the ESQPTs predicted by the collective models (such as the IBM) seem to be located in too high energy domains, where the collective states are already strongly mixed with the others and the low-energy models loose their relevance. However, the critical phenomena for excited states may be available to experimental study in molecular, mesoscopic and quantum optical systems, which are often described up to very high relative excitations by similar models as those applied in nuclei at low energies. As proposed recently [25], ESQPTs in such systems may show up through a strongly anomalous dynamical response to some specific external manipulations.

Acknowledgments

This work was supported by the Czech Science Foundation (202/09/0084) and by the Czech Ministry of Education (MSM 0021620859).

References

[1] Carr L D (editor) 2010, *Understanding Quantum Phase Transitions* (Boca Raton, Taylor & Francis).
[2] Hertz J 1976, *Phys. Rev.* B **14**, 1165
[3] Gilmore R and Feng D H 1978, *Nucl. Phys.* A **301**, 189
Gilmore R 1979, *J. Math. Phys.* **20**, 891
[4] Gilmore R 1981, *Catastrophe Theory for Scientists and Engineers* (New York, Wiley)
[5] Sachdev S 1999, *Quantum Phase Transitions* (Cambridge University Press)
Vojta M 2003, *Rep. Prog. Phys.* **66**, 2069
[6] Dieperink A E L, Scholten O and Iachello F 1980, *Phys. Rev. Lett.* **44**, 1747
Feng D H, Gilmore R and Deans S R 1981, *Phys. Rev.* C **23**, 1254
[7] Casten R F 2009, *Prog. Part. Nucl. Phys.* **62**, 183
Cejnar P and Jolie J 2009, *ibid.* **62**, 210
[8] Cejnar P, Jolie J and Casten R F 2010, *Rev. Mod. Phys.* **82**, 2155
[9] Iachello F and Oss S 2002, *Eur. Phys. J.* D **19**, 307
Dusuel S, Vidal J, Arias J M, Dukelsky J and García-Ramos J E 2005, *Phys. Rev.* C **72**, 011301(R)
Vidal J, Arias J M, Dukelsky J and García-Ramos J E 2006, *Phys. Rev.* C **73**, 054305
Cejnar P and Iachello F 2007, *J. Phys.* A **40**, 581
[10] Iachello F and Arima A 1987, *The Interacting Boson Model* (Cambridge University Press)
[11] Iachello F and Levine R D 1995, *Algebraic Theory of Molecules*, (Oxford University Press)
[12] Frank A and Van Isacker P 2005, *Symmetry Methods in Molecules and Nuclei* (México D.F., S y G editores)
[13] Rowe D J and Wood J L 2010, *Fundamentals of Nuclear Models* (Singapore, World Scientific)
[14] Rowe D J, Bahri C and Wijesundera W 1998, *Phys. Rev. Lett.* **80**, 4394
Iachello F 2000, *ibid.* **85**, 3580
Iachello F 2001, *ibid.* **87**, 052502
Rowe D J 2004, *ibid.* **93**, 122502
[15] Zhang W-M and Feng D H 1995, *Phys. Rep.* **252**, 1
[16] Zhang W-M, Feng D H and Gilmore R 1990, *Rev. Mod. Phys.* **62**, 867
[17] Landau L 1937, *Phys. Z. Sowjet.* **11**, 26; 545
[18] Cejnar P, Macek M, Heinze S, Jolie J and Dobeš J 2006, *J. Phys.* A **39**, L515
[19] Caprio M A, Cejnar P and Iachello F 2008, *Ann. Phys.* (N.Y.) **323**, 1106
[20] Cejnar P and Stránský P 2008, *Phys. Rev.* E **78**, 031130
[21] Leyvraz F and Heiss W D 2005, *Phys. Rev. Lett.* **95**, 050402
[22] Pérez-Bernal F and Iachello F 2008, *Phys. Rev.* A **77**, 032115
[23] Reis M, Terra Cunha M O, Oliviera A C and Nemes M C 2005, *Phys. Lett.* A **344**, 164
[24] Caprio M A, Skrabacz J H and Iachello F 2011, *J. Phys.* A **44**, 075303
[25] Pérez-Fernández P, Cejnar P, Arias J M, Dukelsky J, García-Ramos J E and Relaño A 2011, *Phys. Rev.* A **83**, 033802

Investigating the symmetry energy of nuclear equation of state with heavy-ion reactions.

S Yennello[1,2], Z Kohley[1,2], L W May1[1,2], S Wuenschel[1,2], A Bonasera[2,3], R Tripathi1[1,2], G.A Souliotis[1,2,4]

[1] Chemistry Department, Texas A&M University, College Station, Texas 77843, USA
[2] Cyclotron Institute, Texas A&M University, College Station, Texas 77843, USA
[3] Laboratori Nazionali del Sud, INFN, I-95123 Catania, Italy
[4] Laboratory of Physical Chemistry, Department of Chemistry, National and Kapodistrian University of Athens, Athens, Greece

Email: yennello@comp.tamu.edu

Abstract. Heavy-ion collisions provide a unique opportunity to examine nuclear matter at temperatures, densities, and neutron-to-proton (N/Z) ratios away from that of ground state nuclei. A wide range of observables from heavy-ion collisions has been used to study and constrain the nuclear equation of state (EOS). Recently much work has been focused on the symmetry energy of the EOS. The transverse collective flow of light charged particles and intermediate mass fragments is sensitive to the fragmentation mechanism and – in particular – the density dependence of the nuclear symmetry energy. The present work demonstrates the impact of the N/Z of the system on the transverse flow of LCPs and IMFs from the reactions ^{70}Zn + ^{70}Zn, ^{64}Zn +^{64}Zn, and ^{64}Ni + ^{64}Ni at 35 Mev/nucleon. The IMF flow is consistent with a "stiff" equation of state as implemented in the AMD model.

1. Introduction

Heavy-ion collisions provide a unique opportunity to examine nuclear matter at temperatures, densities, and neutron-to-proton (N/Z) ratios away from that of ground state nuclei. A wide range of observables from heavy-ion collisions have been used to study and constrain the nuclear equation-of-state (EOS) [1, 2]. While the EOS for symmetric nuclear matter (N=Z) is relatively well constrained [3–5], predictions for the density dependence of the symmetry energy, Esym(ρ), can still vary widely [2, 5]. The collective transverse flow of light charged particles (LCPs) has been predicted to be a useful probe for applying constraints on the asymmetric part of the EOS at both high and low densities [2, 6]. The transverse collective flow has been shown to depend on both the mass and N/Z of the colliding system[7-12]. The examination of the balance energy demonstrated that the transverse flow was strongly dependent on the mass, A, of the colliding system [7]. Pak et al. have shown that the transverse collective flow for Z=1-3 particles increases with an increasing neutron-to-proton ratio of the system (N/Z)sys [9,10]. The isospin dependence of the transverse flow and balance energy was attributed to the isospin dependent potential and in-medium NN-cross sections [11,12]. Along with the mass-dependent mechanisms (mean-field and NN-collisions), theoretical simulations have also demonstrated the importance of the Coulomb potential in describing the transverse flow [13-15].

Published under licence by IOP Publishing Ltd

While LCPs have typically been used in studying the transverse flow, we also investigated IMFs to explore the dependences of the transverse flow on the mass and charge of the colliding systems. Additionally, the IMF flow is shown to be viable probe to the EOS, particularly the density dependence of the symmetry energy.

2. Experiment

The reactions were measured using the NIMROD-ISiS detector array [16] at the Texas A&M Cyclotron Institute. Beams of ^{70}Zn, ^{64}Zn, and ^{64}Ni were accelerated by the K500 superconducting cyclotron to 35 MeV/nucleon. These beams were impinged on ^{70}Zn, ^{64}Zn, and ^{64}Ni self-supporting targets, respectively. NIMROD-ISiS consists of 14 concentric rings providing coverage from 3.6 to 167 in lab as shown in figure 1. The first 8 rings, ranging from 3.6 to 45.0, had the same geometry as the INDRA detector [17] and the final 6 rings were of the ISiS geometry [18]. Isotopic resolution was achieved, in the forward angles, for Z=1-17 particles and elemental identification was obtained up through the charge of the beam. In the backward angles detector thresholds allowed only for identification of Z=1-2 particles. The entire charged particle array was surrounded by the 4π TAMU Neutron Ball which provided an average event-by-event neutron multiplicity.

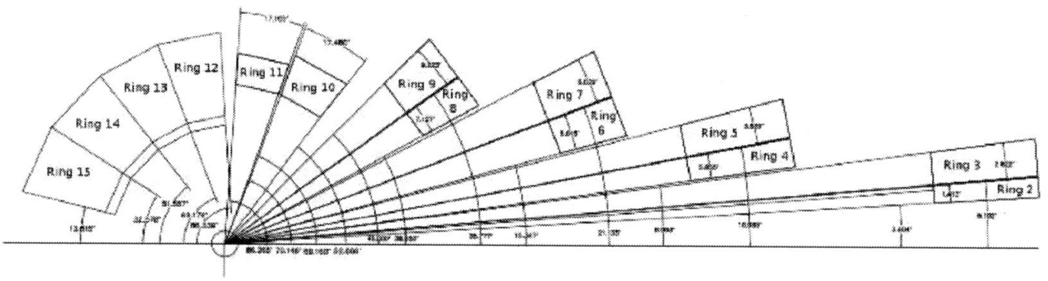

Figure 1. NIMROD-ISiS array

3. Impact parameter determination

The impact parameter for the experimental data was determined using minimum bias 2-D distributions of the raw neutron multiplicity plotted against the charged particle multiplicity for each system. Five bins (0-4) were created from the 2-D distributions such that each bin would represent a b/b_{max}, or b_{red}, width of 0.2 if one assumes a corresponding triangular impact parameter distribution. Molecular dynamics simulations [19,20], filtered using a software replica of the NIMROD-ISiS array, showed that the 80% (20%) of the events in the most peripheral (central) bin were correctly identified. The impact parameter determination for the more central bins was less certain. Thus, Bin 0 does not necessarily contain the most central events but rather the most violent events, while the bins representing the peripheral collisions do provide a relatively accurate impact parameter estimation.

4. Transverse Flow of LCP

In order to calculate the in-plane transverse momentum for the fragments, the reaction plane for each event was reconstructed using the azimuthal correlation method [21]. The azimuthal correlation method does not differentiate the forward, quasi-projectile, and backward, quasi- target, sides of the flow. Therefore, the forward flow side of the reaction plane was determined using the transverse momentum analysis method [22]. The particle of interest (POI) was removed from the calculation of the reaction plane in order to avoid autocorrelations [21-23]. Thus, the reaction plane was calculated for each particle in an event rather than once for the whole event. In order to ensure that only quasi-complete events were used in the analysis an event criterion was imposed such that the total detected charge for an event must be greater than 40% of the total charge in the colliding system. The transverse flow is often quantified as the slope of the average in-plane momentum, <Px/A>, at mid-rapidity as shown in figure 2 for protons. In the following analysis, the flow was defined as the slope of the <Px/A> over a reduced rapidity region (Yr) of -0.45 to 0.45.

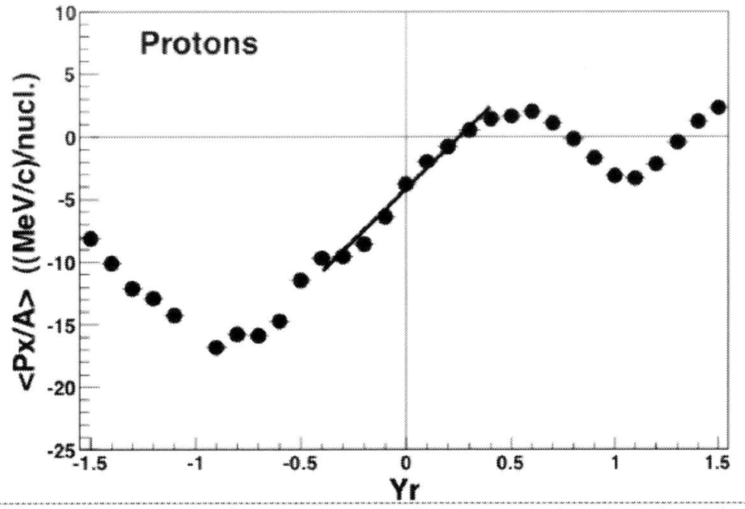

Figure 2 Average in plane momentum of protons as a function of rapidity from the ^{64}Ni+^{64}Ni reaction. The slope of the line at mid-rapidity represents the measured flow.

Figure 3 demonstrates the effect of both the isospin of the reaction system as well as the mass of both the particle of interest and the reaction system. The flow of Z=1 and Z=2 particles are shown as a function of the N/Z of the reaction system. As one can see the flow is larger for the ^{64}Ni + ^{64}Ni system than it is for the ^{64}Zn + ^{64}Zn system. This increase in flow with neutron richness, which is shown by the blue line, is consistent with earlier data by Pak.[9,10]. Based only on the N/Z one would expect that the ^{70}Zn+^{70}Zn would have yet a larger flow. However the flow measured from the ^{70}Zn+^{70}Zn system is lower than that of the ^{64}Ni+^{64}Ni system because it is a larger mass system. This can be understood through the mass dependence of the transverse flow which is related to the balance energy relationship derived by Westfall et al. [7]. Thus, one would expect the ^{70}Zn system to exhibit a decreased flow in comparison to the A=64 systems since it has a lower balance energy due to the increased repulsive NN-collisions relative to the attractive mean-field potential. As expected the flow for the Z=2 particles is larger than for the Z=1 particles.

The effect of the isospin of the particle of interest is shown in figure 4. The observation that the flow increases for more neutron-rich systems made from the nonisotopically resolved data (shown in figure 3) holds true here when one breaks it down into the individual isotopes. For each of the isotopes shown in figure 4 the ^{64}Zn+^{64}Zn system has a lower flow than the ^{64}Ni +^{64}Ni system.

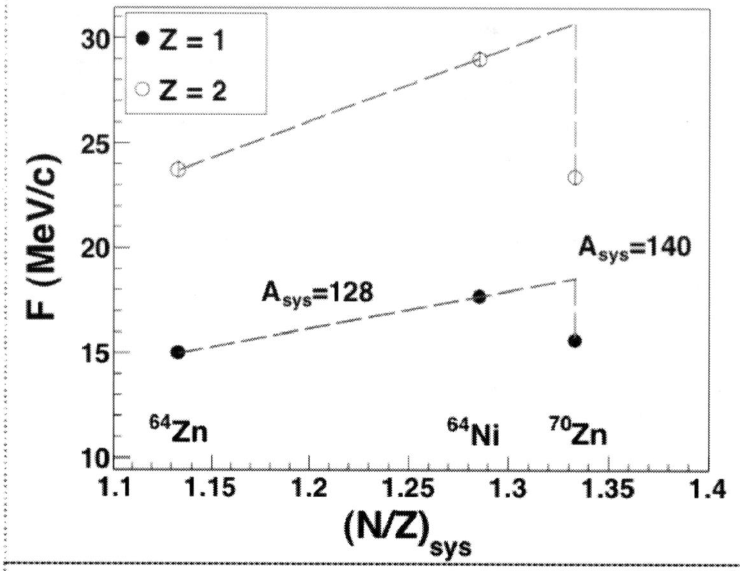

Figure 3. Transverse collective flow of Z=1,2 particles as a function of the N/Z of the reaction. The lines are drawn to guide the eye. The blue line represents the expected change in the flow for a constant mass $A_{proj}=A_{tgt}=64$ system. The red line represents the decrease in flow due to the mass for the $A_{proj}=A_{tgt}=70$ system based upon the blue line for the A=64 system.

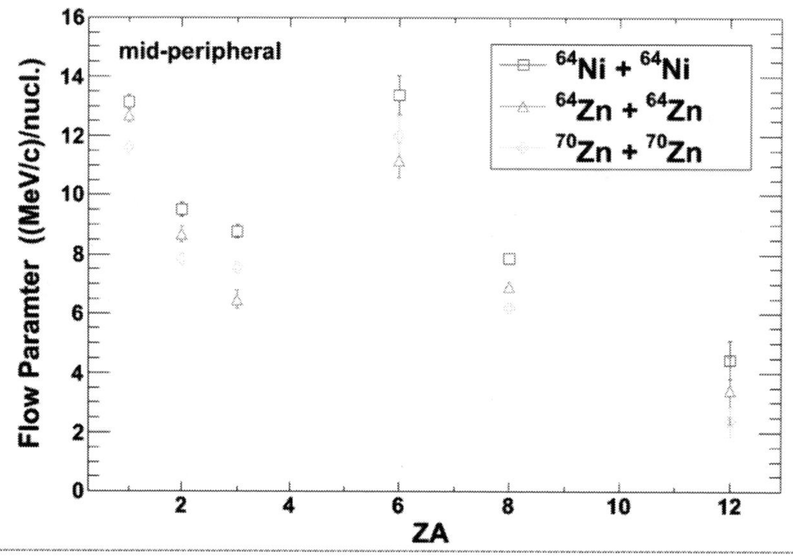

Figure 4. Transverse collective flow per nucleon as a function of the charge times the mass of the measured particle for the ^{70}Zn + ^{70}Zn, ^{64}Zn + ^{64}Zn, and ^{64}Ni + ^{64}Ni reactions at 35 Mev/nucleon.

Figure 4 plots the flow per nucleon as a function of the Z times the A of the particle. Using the Flow/nucleon allows us to remove the difference in flow due to particle size and focus on more interesting differences. Within a given Z the flow always decreases as the N/Z of the fragment increases. The triton to ^{3}He comparison demonstrated the effect of the neutron richness of the fragment where there is not difference in size.

5. IMF flow

In the NIMROD-ISiS array thresholds produced incomplete detection of intermediate mass fragments (IMFs) at negative reduced rapidities ($Y_r=Y_{cm}/Y_{cm,proj}$). Therefore, the transverse flow was quantified by calculating the average in-plane transverse momentum from $0.0<Y_r<0.45$ [24–26]. The flow is extracted only from the positive rapidity fragments and is designated as $\overline{\langle P_x \rangle}$. Details of this method can be found in Ref. [27].

The transverse flow $\overline{\langle P_x \rangle}$, for Z=1-9 particles is shown in Figure 5 for the five centrality bins, ranging from Bin 0 (most violent collisions) to bin 4 (most peripheral collisions). The expected increase in the transverse flow with the increasing charge of the fragments is clear [9, 28]. In bin 0 the $\overline{\langle P_x \rangle}$ of the IMFs from the ^{64}Ni and ^{64}Zn systems are nearly equivalent and larger than the ^{70}Zn system. This is consistent with previous work by Westfall on the mass dependence of the balance energy. [7] In the peripheral reactions, Bins 3 and 4, the $\overline{\langle P_x \rangle}$ of the IMFs from the Zn systems become similar and decreased with respect to the ^{64}Ni system. This represents a clear dependence of the charge of the system on the IMF flow. The larger repulsive Coulomb force in the Zn (Z=30) systems causes a decreased flow in comparison to the ^{64}Ni (Z=28) system. The increased effect of the charge-dependent forces, relative to the mass-dependent forces, in the peripheral collisions may be due to the decreased interaction volume. For example, the number of NN-collisions would be greatly diminished in the peripheral reactions. A separation of the IMF flow between all 3 systems occurs in the mid-peripheral reactions, bin 2, in which the IMF's $\overline{\langle P_x \rangle}$ from the ^{64}Zn system is less than that from the ^{64}Ni system yet larger than the ^{70}Zn flow, exhibiting a behavior between the extremes of the mass (bin 0) and charge (bin 4) dependent flow. The difference between the IMF flow in the ^{64}Ni and ^{64}Zn systems is similar to the $(N/Z)_{sys}$ dependence observed by Pak et al. for LCPs in A=58 systems [9]. However, in context with the results from the IMF flow of the ^{70}Zn system, which has a similar $(N/Z)_{sys}$ to the ^{64}Ni system, the difference between the A=64 IMF flow appears to be due to a balancing between the mass and charge dependent mechanisms.

Journal of Physics: Conference Series **322** (2011) 012013 doi:10.1088/1742-6596/322/1/012013

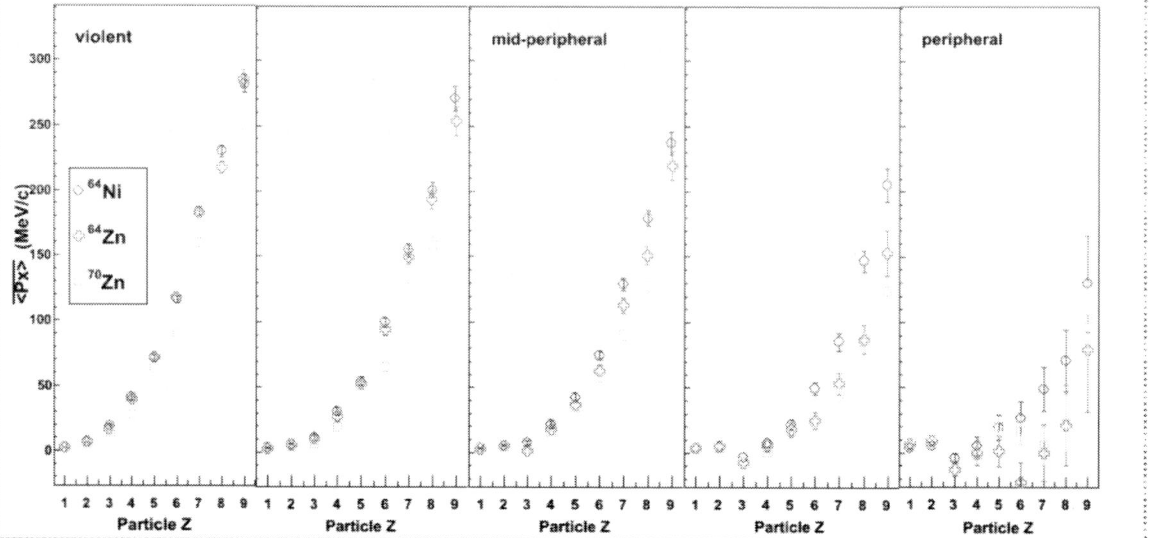

Figure 5. Transverse collective flow of fragments from Z=1-9 as a function of particle Z for five bins in impact parameter. Bin 0 is the most central or most violent while Bin 4 represented the most peripheral collisions.

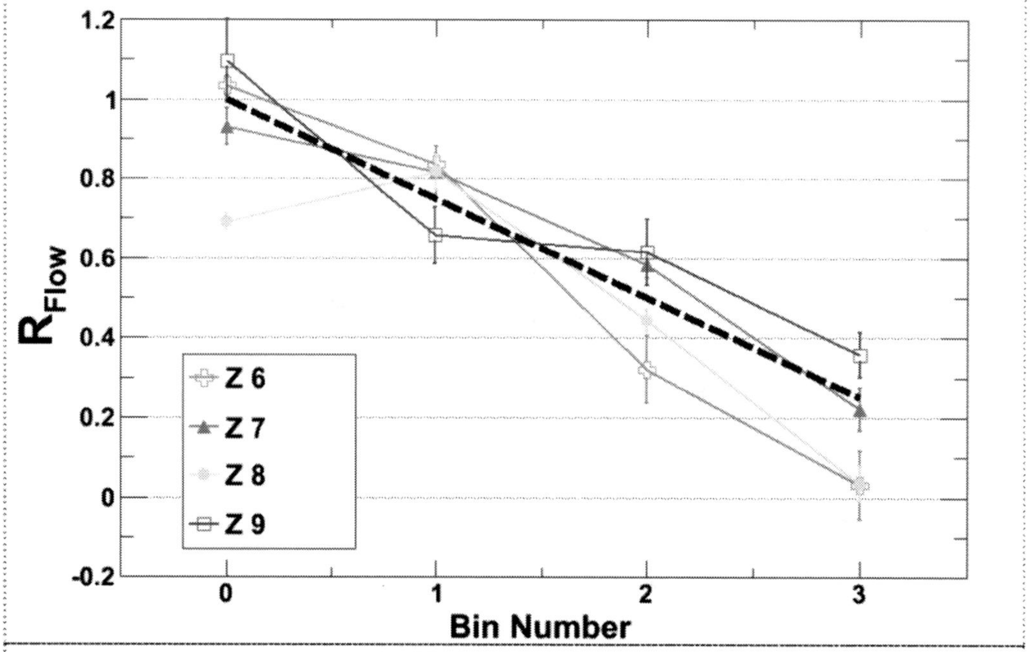

Figure 6. R_{Flow} for Z=6-9 (colored lines) as a function of impact parameter in b_{red} bins. The black dashed line represents a linear transition from R_{Flow}=1 at bin 0 to R_{Flow}=0 at bin 4 (not shown due to statistical uncertainties in bin 4).

In order to examine this trend more quantitatively the ratio

$$R_{Flow} = \frac{\overline{\langle P_x \rangle}_{^{64}Zn} - \overline{\langle P_x \rangle}_{^{70}Zn}}{\overline{\langle P_x \rangle}_{^{64}Ni} - \overline{\langle P_x \rangle}_{^{70}Zn}}$$

can be used to define the magnitude of the flow from the ^{64}Zn system in comparison to the ^{64}Ni and ^{70}Zn systems. Thus, when $R_{Flow}=1$ the IMF flow of the ^{64}Zn system equals that of the ^{64}Ni system and when $R_{Flow}=0$ the ^{64}Zn and ^{70}Zn systems have equivalent values of flow. In Figure 6, the individual R_{Flow} values of the Z=6-9 fragments and the average R_{Flow} value of Z=4-9 fragments are plotted as a function of the centrality bin number. The ratio values exhibit a systematic trend from $R_{Flow} = 1$ for the most violent collisions to $R_{Flow} = 0$ for the most peripheral reactions. This trend, observed in Figures 5 and 6, shows a transition from the IMF's $\langle P_x \rangle$ being strongly dependent on the mass of the system to a dependence on the charge of the system.

The observed mass to charge dependence of the IMF transverse flow should be sensitive to the density dependence of the symmetry energy since there is a mean-field component to the flow. Scalone et al. predicted that the difference in the transverse flow of LCPs from two systems with the same mass and differing $(N/Z)_{sys}$ would be sensitive to Esym(ρ) [29]. Therefore, changing the isospin-dependent part of the mean-field should affect the balance between the mass and charge dependent forces. The antisymmetrized molecular dynamics with wave packet Diffusion and Shrinking (AMD-DS) model [20] was used to investigate the sensitivity of the IMF flow to Esym(ρ). The dynamics of the reaction were simulated up to a time of 300 fm/c, after which the GEMINI code [30] was used to statistically de-excite the hot fragments. The excitation energy and angular momentum of the hot fragments, produced by AMD, were used as inputs into the GEMINI code. Additional details about the coupling of dynamical AMD and statistical GEMINI codes can be found in Ref. [20]. The momentum dependent Gogny and Gogny-AS effective interactions provided an incompressibility of symmetric nuclear matter of K=228 MeV while allowing for the density dependence of the symmetry energy to be varied [20]. The Gogny and Gogny-AS interactions produce a soft and stiff density dependence of the symmetry energy, respectively [31].

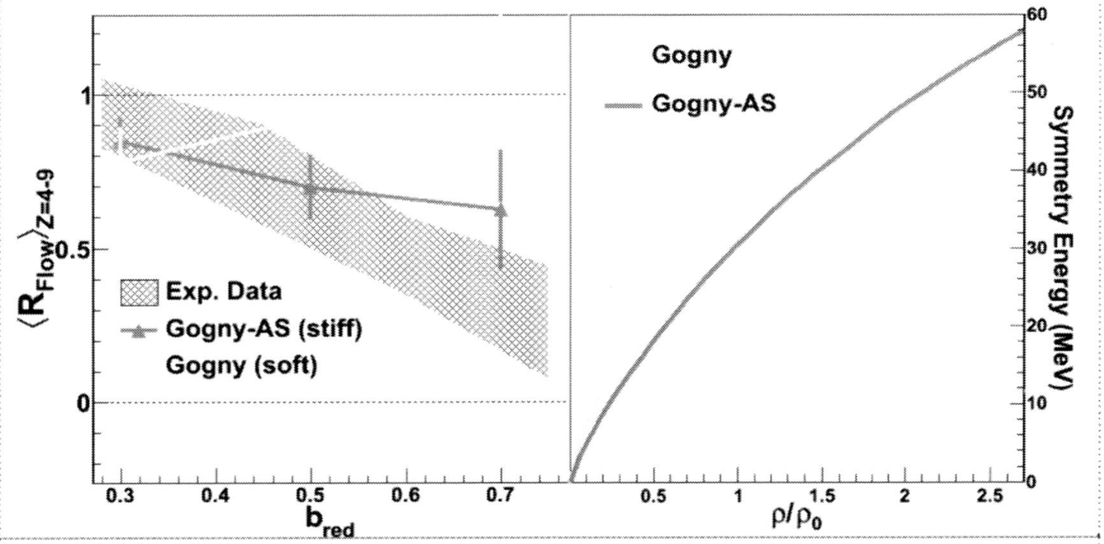

Figure 7. The ratio of the difference in transverse collective flow between the ^{64}Zn and ^{64}Ni compared to that of the ^{70}Zn system for Z=4-9 fragments as a function of the reduced impact parameter (left). The corresponding symmetry energy as a function of density for the interactions in the AMD results (right).

In Figure 7 the average R_{Flow} value for Z=4-9 fragments is shown as a function of b_{red} from the AMD-Gemini simulations in comparison to the experimental data. The experimental results are equivalent to those presented in Figure 6 except that R_{Flow} is shown as a function of the average bred of each centrality bin. The average b_{red} was determined using the filtered molecular dynamics simulations to provide an estimate of the impact parameter range selected in each centrality bin. The impact parameter for each event of the AMD-Gemini simulation was known and, therefore, the average b_{red} values shown in Figure 7 are exact. While the same experimental procedure discussed above was used to extract the IMF flow, the AMD-Gemini results (Figure 7) were not filtered due to statistical limitations. The results of Figure 7 demonstrate that the differences in the IMF flow between systems have a strong sensitivity to the density dependence of the symmetry energy. The Gogny-AS interaction, or stiff Esym(ρ), clearly demonstrates the best agreement with the experimental data, showing a decreasing $\langle R_{Flow} \rangle_{Z=4-9}$ value with increasing b_{red}. In comparison, the soft symmetry energy parameterization, or Gogny interaction, is unable to reproduce the experimental trend. In the Gogny calculation the ^{64}Zn flow increases relative to the ^{64}Ni flow, eventually becoming larger ($R_{Flow} > 1$). This is related to the larger symmetry energy at low-density for the Gogny interaction, which is more repulsive for the more neutron-rich ^{64}Ni system relative to the ^{64}Zn system. The isospin-dependent part of the Gogny-AS interaction is less repulsive at low-density and therefore, the ^{64}Ni flow remains larger than the ^{64}Zn flow producing agreement with the experimental data. It is clear that the isospin-dependent part of the interaction is an important component in describing the observed transition from a mass to charge dependence of the IMF transverse flow.

6. Conclusions

The transverse flow of the IMFs has been shown to be sensitive to both the mass and charge of the colliding system. The strong dependence of the IMF flow on the charge of the system in the peripheral collisions provides new insight into the mechanims responsible for the transverse flow. The AMD-Gemini simulation demonstrated that the differences in the IMF flow between the systems was sensitive to the isospin-dependent part of the nucleon-nucleon interaction. Comparison between the experiment and AMD-Gemini provided strong evidence supporting a stiff Esym(ρ). Future research examining the IMF flow should allow for additional constraints on the nuclear Equation of State.

Acknowledgements

We would like to thank the staff members of the Texas A&M Cyclotron Institute for the excellent beam quality. This work was supported in part by the Robert A. Welch Foundation through grant No. A-1266, and the Department of Energy through grant No. DE-FG03-93ER40773. We would further like to thank Dr. A. Ono for use of the AMD code and Dr. R. J. Charity for use of the Gemini code.

References

[1] Baran V, Colonna M, Greco V and Di Toro M 2005 *Phys. Rep.* **410** 335
[2] Li B A, Chen L W and Ko C M 2008 *Phys. Rep.* **464** 113
[3] Danielewicz P, Lacey R and Lynch W G 2002 *Science* **298** 1592
[4] Shlomo S, Kolomietz V M and Colo G 2006 *Eur. Phys. J.* A **30** 23
[5] Fuchs C and Wolter H H 2006 *Eur. Phys. J.* A **30** 5
[6] Di Toro M, Yennello S J and Li B A 2006 *Eur. Phys. J.* A **30** 153
[7] Westfall G D, Bauer W, Craig D *et al* 1993 *Phys. Rev. Lett.* **71** 1986
[8] Li B A, Ko C M and Bauer W 1998 *Int. J. Mod. Phys.* E **7** 147
[9] Pak R *et al* 1997 *Phys. Rev. Lett.* **78** 1022
[10] Pak R *et al* 1997 *Phys. Rev. Lett.* **78** 1026
[11] Liewen C, Fengshou Z and Genming J 1998 *Phys. Rev.* C **58** 2283
[12] Li B A, Ren Z, Ko C M and Yennello S J 1996 *Phys. Rev. Lett.* **76** 4492
[13] Soff S, Bass S A, Hartnack C, Stocker H and Greiner W 1995 *Phys. Rev.* C **51** 3320

[14] Magestro D J, Bauer W, Bjarki O *et al* 2000 *Phys. Rev.* C **61** 021602(R)
[15] de la Mota V, Sebille F, Farine M, Remaud B and Schuck P 1992 *Phys. Rev.* C **46** 677
[16] Wuenschel S *et al* 2009 *Nucl. Instrum. Methods Phys. Res.* A **604** 578
[17] Pouthas J *et al* 1995 *Nucl. Instrum. Methods Phys. Res.* A **357** 418
[18] Kwiatkowski K *et al* 1995 *Nucl. Instrum. Methods Phys. Res.* A **360** 571
[19] Papa M, Giuliani G and Bonasera A 2005 *J. Comput. Phys.* **208** 403
[20] Ono A and Horiuchi H 2004 *Prog. Part. Nucl. Phys.* **53** 501
[21] Wilson W K, Lacey R, Ogilvie C A and Westfall G D 1992 *Phys. Rev.* C **45** 738
[22] Danielewicz P and Odyniec G 1985 *Phys. Lett.* B **157** 146
[23] Cussol D *et al* 2002 *Phys. Rev.* C **65** 44604
[24] Ogilvie C A *et al* 1989 *Phys. Rev.* C **40** 2592
[25] Ono A and Horiuchi H 1995 *Phys. Rev.* C **51** 299
[26] Bonasera A, Gulminelli F and Molitoris J 1994 *Phys. Rep.* **243** 1
[27] Kohley Z *et al* 2010 *Phys. Rev.* C **82** 064601
[28] Reisdorf W and Ritter H G 1997 *Annu. Rev. Nucl. Part. Sci. 47* 663; Huang M J, Lemmon R C, Daffin F, Lynch W G *et al* 1996 *Phys. Rev. Lett.* **77** 3739
[29] Scalone L, Colonna M and Di Toro M 1999 *Phys. Lett.* B **461** 9
[30] Charity R J, McMahan M A, Wozniak G J *et al* 1988 *Nucl. Phys.* A **483** 371
[31] Ono A 2002 *Prog. Theor. Phys. Supp.* **146** 378

Journal of Physics: Conference Series **322** (2011) 012014 doi:10.1088/1742-6596/322/1/012014

Spin and flavor content of octet baryons

Roelof Bijker[1] **and Elena Santopinto**[2]

[1] Instituto de Ciencias Nuclares, Universidad Nacional Autónoma de México, AP 70-543, 04510 Mexico DF, Mexico

[2] I.N.F.N., Sezione di Genova, via Dodecaneso 33, 16164 Genova, Italy

E-mail: bijker@nucleares.unam.mx

Abstract. We discuss the spin and flavor content of the proton in an unquenched quark model. It is shown that the inclusion of hadron loops leads automatically to an excess of $\bar{d}$ over $\bar{u}$ and introduces a sizeable contribution of orbital angular momentum to the spin of the proton. Whereas the flavor asymmetry and the orbital angular momentum are dominated by pion loops, the contribution of the sea quark spins to the proton spin arises almost entirely from (excited) vector meson loops.

1. Introduction

In the constituent quark model (CQM) hadrons are described in terms of system of constituent (or valence) quarks and antiquarks, qqq for baryons and $q\bar{q}$ for mesons. Despite the success of the quark model there is strong evidence for the existence of exotic degrees of freedom (other than valence quarks) in hadrons, in particular for the need to include the effects of quark-antiquark pairs (hadron loops). The importance of these quark-antiquark configurations (or higher Fock components in baryon wave functions) is evident from measurements of the $\bar{d}/\bar{u}$ asymmetry in the nucleon sea [1], parity-violating electron scattering experiments [2, 3], the proton spin crisis [4], as well as from analysis of helicity amplitudes [5] and strong couplings of baryon resonances [6, 7].

The aim of this contribution is to study the spin and flavor content of the proton in an unquenched quark model in which the effects of quark-antiquark pair creation ($u\bar{u}$, $d\bar{d}$ and $s\bar{s}$) are taken into account in an explicit form via a $^3P_0$ coupling mechanism [8, 9, 10]. The numerical results for the flavor asymmetry [9] and the proton spin [8] are analyzed by means of a simple exactly solvable meson-cloud model. Finally, we show some predictions for other octet baryons as well.

2. Unquenched quark model

The impact of $q\bar{q}$ pairs in hadron spectroscopy was originally studied by Törnqvist and Zenczykowski in a quark model extended by the $^3P_0$ model [11]. Subsequently, the effects of hadron loops in mesons was studied by Geiger and Isgur in a flux-tube breaking model in which the $q\bar{q}$ pairs are created in the $^3P_0$ state with the quantum numbers of the vacuum [12, 13]. It was shown that cancellations between apparently uncorrelated sets of intermediate states occur in such a way that the modification in the linear potential can be reabsorbed, after renormalization, in the new strength of the linear potential [12]. In addition, the quark-antiquark pairs do not destroy the good CQM results for the mesons [12] and preserve the OZI hierarchy [13] provided

Figure 1. One-loop diagram at the quark level.

that the sum be carried out over a large tower of intermediate states. The basic idea of the importance to carry out a sum over a complete set of intermediate states was already contained in [11]. An extension to baryons was presented in [14] in which the effects of $s\bar{s}$ loops in the proton were studied combining harmonic oscillator wave functions for baryons and mesons and a ${}^3P_0$ pair creation mechanism.

The present approach is motivated by earlier studies on extensions of the quark model that employ a ${}^3P_0$ model for the $q\bar{q}$ pair creation [11, 14]. Our approach is based on a CQM to which the quark-antiquark pairs with vacuum quantum numbers are added as a perturbation [8, 9, 14]. The pair-creation mechanism is inserted at the quark level and the one-loop diagrams are calculated by summing over a complete set of intermediate baryon-meson states (BC in Fig. 1). Under these assumptions, the baryon wave function consists of a zeroth order three-quark configuration $\mid A\rangle$ plus a sum over all possible higher Fock components due to the creation of ${}^3P_0$ quark-antiquark pairs

$$\mid \psi_A\rangle = \mathcal{N}\left[\mid A\rangle + \sum_{BClJ}\int d\vec{k}\mid BC\vec{k}\,lJ\rangle\frac{\langle BC\vec{k}\,lJ\mid T^\dagger\mid A\rangle}{M_A - E_B - E_C}\right] . \qquad (1)$$

Here A denotes the initial baryon, B and C represent the intermediate baryon and meson, and M_A, E_B and E_C are their respective energies, $\vec{k}$ and l the relative radial momentum and orbital angular momentum of B and C, and J is the total angular momentum $\vec{J} = \vec{J}_B + \vec{J}_C + \vec{l}$. The operator $T^\dagger$ creates a quark-antiquark pair in the ${}^3P_0$ state with the quantum numbers of the vacuum: $L = S = 1$ and $J = 0$, color singlet and flavor singlet [8, 9, 15]. Therefore, the $SU(3)$ flavor symmetry of the valence quark configuration $\mid A\rangle$ is broken by the quark-antiquark pairs via the energy denominator, but the $SU(2)$ isospin symmetry is still preserved. In the special case of the closure limit in which the energy denominator of Eq. (1) is a constant, the flavor symmetry of the valence quark configuration is recovered.

In order to calculate the effects of quark-antiquark pairs on an observable, one has to evaluate the contribution of all possible intermediate states. By using a combination of group theoretical and computational techniques, the sum over intermediate states is carried out up to saturation and not only for the first few shells as in previous studies [11, 14]. In addition, the contributions of quark-antiquark pairs can be evaluated for any initial baryon (ground state or resonance) and for any flavor of the $q\bar{q}$ pair (not only $s\bar{s}$ as in [14], but also $u\bar{u}$ and $d\bar{d}$), and for any model of baryons and mesons, as long as their wave functions are expressed in the basis of harmonic oscillator wave functions [8, 9].

The unquenching of the quark model has to be done in such a way as to maintain the phenomenological successes of the CQM. In applications to mesons, it was shown that the inclusion of quark-antiquark pairs does not destroy the good CQM results [12] and preserves

Table 1. Contributions to the flavor asymmetry of the proton [9]. $N^2 = 1 + a^2 + b^2 + c^2 + d^2$ is the normalization factor of the wave function of Eq. (3).

	Unquenched QM		Meson-Cloud
	0-4 $\hbar\omega$	0 $\hbar\omega$	Eq. (4)
$N\pi$	0.195	0.177	$2a^2/3N^2$
$\Delta\pi$	-0.016	-0.010	$-b^2/3N^2$
$N\pi\eta_8\eta_1$	-0.028	-0.018	$-2a(c + d\sqrt{2})/3N^2$
$N\rho$	0.050	0.012	
$\Delta\rho$	-0.017	-0.003	
$N\rho\omega_8\omega_1$	-0.033	-0.010	
Total	0.151	0.147	

the OZI hierarchy [13]. In a similar fashion, we showed that the CQM results for the magnetic moments of the octet baryons also hold in the unquenched constituent quark model (UCQM) [8].

In this contribution, we use harmonic oscillator wave functions up to five oscillator shells for the intermediate baryons and mesons. All parameters were taken from the literature without attempting to optimize their values in order to improve the agreement with experimental data [8, 9].

3. Flavor content

The flavor asymmetry of the proton $\mathcal{A}(p)$ is related to the Gottfried integral S_G for the difference of the proton and neutron electromagnetic structure functions as

$$S_G = \int_0^1 \frac{F_2^p(x) - F_2^n(x)}{x} dx = \frac{1}{3} - \frac{2}{3}\int_0^1 \left[\bar{d}_p(x) - \bar{u}_p(x)\right] dx = \frac{1}{3}[1 - 2\mathcal{A}(p)] . \qquad (2)$$

Under the assumption of a flavor symmetric (or rather flavor independent) sea one obtains the Gottfried sum rule $S_G = 1/3$ [16, 1], whereas any deviation from this value is an indication of the $\bar{d}/\bar{u}$ asymmetry of the nucleon sea, thus providing evidence of the existence of higher Fock components (such as $qqq - q\bar{q}$ configurations) in the proton wave function. The first clear evidence of a violation of the Gottfried sum rule came from the New Muon Collaboration (NMC) [17] which was later confirmed by Drell-Yan experiments [18, 19] and a measurement of semi-inclusive deep-inelastic scattering [20]. All experiments show evidence that there are more $\bar{d}$ quarks in the proton than there are $\bar{u}$ quarks [1]. The final NMC value is 0.2281 ± 0.0065 at $Q^2 = 4$ (GeV/c)^2 for the Gottfried integral over the range $0.004 \leq x \leq 0.8$ [17], which implies a flavor asymmetric sea. The observed flavor asymmetry is far too large to be accounted for by processes that can be described by QCD in perturbative regime and therefore has to attributed to non-perturbative QCD mechanisms. It was shown in the framework of the meson-cloud model, that the coupling of the nucleon to the pion cloud provides a mechanism that is able to produce a flavor asymmetry due to the dominance of $n\pi^+$ among the virtual configurations [21].

In the unquenched quark model, the flavor asymmetry of the proton can be calculated directly from the difference of the number of $\bar{d}$ and $\bar{u}$ sea quarks in the proton, even in the absence of explicit information on the (anti)quark distribution functions. Table 1 shows that the flavor asymmetry for the proton in the UCQM is 0.151 which corresponds to a value of the Gottfried integral of 0.232, remarkably close to the experimental value. The main contribution to the flavor asymmetry of the proton is due to the pion loops, especially the $n\pi^+$ intermediate state, thus confirming in an explicit calculation the explanation given in Refs. [21] in the context of the meson-cloud model. In addition, we find that there are important contributions from the $\Delta\pi$ channel and, especially, from the off-diagonal terms $p\pi^0$-$p\eta_8$ and $p\pi^0$-$p\eta_1$ which together are of the order of 15-20 % of that of the $N\pi$ channel, but with the opposite sign (see Table 1). The contribution of the intermediate vector mesons is very small due to a cancelation between the $n\rho^+$ and the $\Delta\rho$ channels and the cross terms $p\rho^0$-$p\omega_8$ and $p\rho^0$-$p\omega_1$. Kaon loops do not contribute to the proton flavor asymmetry. Table 1 shows that the full four-shell calculation is dominated by the contribution of the ground state intermediate baryons and mesons (0 $\hbar\omega$). Both columns show the same qualitative behavior: dominance of the pion loops with a small negative correction of the order of 10-15 % due to the off-diagonal terms involving π and η pseudoscalar mesons and an almost vanishing contribution from the vector mesons.

A similar result can be obtained in the meson-cloud model by considering a proton wave function including not only pion loops but also eta loops

$$\begin{aligned} |\psi_p\rangle \quad \rightarrow \quad & |p\rangle + a\left[\frac{1}{\sqrt{3}}\,|p\pi^0\rangle - \sqrt{\frac{2}{3}}\,|n\pi^+\rangle\right] + c\,|N\eta_8\rangle + d\,|N\eta_1\rangle \\ & +b\left[\frac{1}{\sqrt{2}}\,|\Delta^{++}\pi^-\rangle - \frac{1}{\sqrt{3}}\,|\Delta^+\pi^0\rangle + \frac{1}{\sqrt{6}}\,|\Delta^0\pi^+\rangle\right] . \end{aligned} \tag{3}$$

In the last column of Table 1 we show the contributions of the different terms of Eq. (3) to the flavor asymmetry of the proton to obtain

$$\mathcal{A}(p) = \frac{2a^2 - b^2 - 2a(c + d\sqrt{2})}{3(1 + a^2 + b^2 + c^2 + d^2)} . \tag{4}$$

Since the unquenched quark model is valid not only for the proton, but for all baryons (ground state or resonance), it is straightforward to calculate the flavor asymmetries of the other octet baryons. For the Σ^+ hyperon and the Ξ^0 cascade particle we find $\mathcal{A}(\Sigma^+) = 0.126$ and $\mathcal{A}(\Xi^0) = -0.001$ [9], respectively. The flavor asymmetries of the remaining octet baryons can be obtained by using the isospin symmetry of the unquenched quark model [9]. For example, the excess of $\bar{d}$ over $\bar{u}$ in the proton is related to the excess of $\bar{u}$ over $\bar{d}$ in the neutron, $\mathcal{A}(p) = -\mathcal{A}(n)$. Similar relations hold for the other octet baryons: $\mathcal{A}(\Sigma^+) = -\mathcal{A}(\Sigma^-)$, $\mathcal{A}(\Xi^0) = -\mathcal{A}(\Xi^-)$ and $\mathcal{A}(\Lambda) = \mathcal{A}(\Sigma^0) = 0$. Just as for the proton, the flavor asymmetry of the other octet baryons is expected to be dominated by pion loops, whereas the other contributions are suppressed by the energy denominator in Eq. (1). For the Σ hyperon this is indeed the case, but for the cascade particles the pion loops are suppressed by the value of the $SU(3)$ flavor coupling which is a factor of 5 smaller than that for the proton. Hence for the Ξ hyperons there is no dominant contribution. Since for the Ξ hyperon all contributions are roughly of the same order and small, and moreover some with a positive and others with a negative sign, the value of the flavor asymmetry of the cascade particles is calculated to be small [9].

In Table 2, we show a comparison of some predictions for the flavor asymmetry of the Σ^+ and Ξ^0 hyperons relative to that of the proton. In the unquenched quark model, the flavor asymmetry of the proton is predicted to be of the same order as that of the Σ^+ hyperon and much larger than that of the cascade particle

$$\mathcal{A}(p) \sim \mathcal{A}(\Sigma^+) \gg |\mathcal{A}(\Xi^0)| . \tag{5}$$

This behavior is very different from that obtained in the chiral quark model $\mathcal{A}(\Sigma^+) = 2\mathcal{A}(p) = 2\mathcal{A}(\Xi^0)$ [22], the balance model $\mathcal{A}(\Sigma^+) > \mathcal{A}(\Xi^0) > \mathcal{A}(p)$ [23], and the octet model $\mathcal{A}(p) > |\mathcal{A}(\Xi^0)| > \mathcal{A}(\Sigma^+)$ [24]. The values for the chiral quark model and the balance model were taken from [25].

Table 2. Relative flavor asymmetries of octet baryons.

Model	$\mathcal{A}(\Sigma^+)/\mathcal{A}(p)$	$\mathcal{A}(\Xi^0)/\mathcal{A}(p)$	Ref.
Unquenched CQM	0.833	−0.005	[9]
Chiral QM	2	1	[22]
Balance Model	3.083	2.075	[23]
Octet couplings	0.353	−0.647	[24]

In order to distinguish between the predictions of the different models and to obtain a better understanding of the non-perturbative structure of QCD, new experiments are needed to measure the flavor asymmetry of hyperons. In particular, the flavor asymmetry of charged Σ hyperons can obtained from Drell-Yan experiments using charged hyperon beams on the proton [24] or by means of backward $K^{\pm}$ electroproduction [26].

4. Spin content

The contribution of the quark spins to the spin of the proton can be obtained from the proton spin structure function g_1^p in combination with the neutron and hyperon semileptonic decays [4]. The observation by the European Muon Collaboration that the total quark spin constitutes only a small fraction of the spin of the nucleon [27] sparked an enormous interest in the spin structure of the proton [4]. Recent experiments show that approximately one third of the proton spin is carried by quarks [28, 29], and that the gluon contribution is rather small (either positive or negative) and compatible with zero [30]. This rules out the possibility that most of the missing spin be carried by the gluon and indicates that the origen of the missing spin of the proton has to be attributed to other mechanisms.

In the unquenched quark model, the effect of hadron loops on the fraction of the proton spin carried by the quark (antiquark) spins and orbital angular momentum can be studied in an explicit way [8]. As in other effective models [4], gluonic effects associated with the axial anomaly are not included, and therefore the contribution from the gluons is missing from the outset. The total spin of the proton can then be written as the sum of the contributions from the quark (and antiquark) spins and orbital angular momentum

$$1 = 2\Delta J = \Delta\Sigma + 2\Delta L \; . \tag{6}$$

Table 3 shows that the inclusion of the quark-antiquark pairs has a dramatic effect on the spin content of the proton. Whereas in the CQM the proton spin is carried entirely by the (valence) quarks, in the unquenched calculation the contributions of the valence quark spins, the sea quark spins and the orbital angular momentum to the proton spin are comparable in size and equal to approximately 38, 30 and 32 %, respectively. The importance of orbital angular momentum to the proton spin was discussed many years ago by Sehgal [31] and Ratcliffe [32] in the context of the quark-parton model and, more recently, by Myhrer and Thomas in framework of the bag model [33].

These results can be understood in a qualitative way by considering the proton wave function of Eq. (3), whose contribution to the orbital angular momentum can be derived as

$$\Delta L = \frac{2a^2 - b^2 + 2c^2 + 2d^2}{3(1 + a^2 + b^2 + c^2 + d^2)} \; . \tag{7}$$

The effects of pion loops for the proton flavor asymmetry and the contribution of orbital angular momentum to the proton spin are identical $\mathcal{A}(p) = \Delta L$ as a consequence of the spin and isospin

Table 3. Contribution of quark spins $\Delta\Sigma$ and orbital angular momentum ΔL to the spin of the proton and the Λ hyperon.

		CQM	Unquenched QM		
			Valence	Sea	Total
p	$\Delta\Sigma$	1	0.378	0.298	0.676
	$2\Delta L$	0	0.000	0.324	0.324
	$2\Delta J$	1	0.378	0.622	1.000
Λ	$\Delta\Sigma$	1	0.422	0.429	0.851
	$2\Delta L$	0	0.000	0.149	0.149
	$2\Delta J$	1	0.422	0.578	1.000

properties [34]. Since in the unquenched calculations both the flavor asymmetry and the orbital angular momentum are dominated by pion loops (see Tables 1 and 4), this relation is to a good approximation still valid in the UCQM, $\mathcal{A}(p) = 0.151$ and $\Delta L = 0.162$, respectively. Table 4 shows that, just as for the flavor asymmetry, the orbital angular momentum is dominated by the contribution of the ground state intermediate baryons and mesons ($0\ \hbar\omega$) and in particular by the $N\pi$ channel. The contributions of the eta and kaon loops to the orbital angular momentum, as well as that of the vector mesons, are small with respect to that of the pions.

The situation for the quark spins is completely different. In the unquenched calculations, the contributions of valence and sea quarks are given by $\Delta\Sigma_{\rm val} = 0.378$ and $\Delta\Sigma_{\rm sea} = 0.298$, respectively. While the orbital angular momentum arises almost entirely from the $N\pi$ channel, the sea quark spins are dominated by the intermediate vector mesons with a very small contribution from the pseudoscalar mesons (see second column of Table 5). The third column of Table 5 shows that the contribution of the ground state intermediate baryons and mesons ($0\ \hbar\omega$) is small for both the pseudoscalar and vector mesons, whereas the full calculation is dominated by the contribution of the vector mesons. This shows that for the sea quark spins it is crucial to include the effects of the excited vector mesons which makes the convergence of the sum over intermediate states is much slower. Therefore, the sum was carried out over five complete oscillator shells for both the intermediate baryons and mesons [8].

In a simple meson-cloud model based on ground state baryons and mesons only, the contribution of the sea quark spins is very small. For example, for the proton wave function of Eq. (3) in which only pion and eta loops are taken into account, the contribution of the quark spins is given by $\Delta\Sigma = \Delta\Sigma_{\rm val} + \Delta\Sigma_{\rm sea} = 1 - 2\Delta L$ with

$$\begin{aligned} \Delta\Sigma_{\rm val} &= \frac{1}{1+a^2+b^2+c^2+d^2} , \\ \Delta\Sigma_{\rm sea} &= \frac{-a^2+5b^2-c^2-d^2}{3(1+a^2+b^2+c^2+d^2)} . \end{aligned} \tag{8}$$

With the values of the coefficients a, b, c and d, as determined by comparing the third and fourth columns of Tables 4 and 5, we find that for this case the contribution of the sea quark spins is very small $\Delta\Sigma_{\rm sea} = -0.035$.

The experimental data on the spin structure of the proton have raised many questions about the contributions of valence and sea quarks, gluons and orbital angular momentum to the proton spin. In this respect it is of interest to investigate the spin structure of other octet baryons, in particular the Λ hyperon. In most studies, additional assumptions had to be made about the sea quarks in order to get an estimate of its spin content. For example, the assumption that both

Journal of Physics: Conference Series **322** (2011) 012014 doi:10.1088/1742-6596/322/1/012014

Table 4. Contribution of the orbital angular momentum $2\Delta L$ to the spin of the proton. $N^2 = 1 + a^2 + b^2 + c^2 + d^2$ is the normalization factor of the wave function of Eq. (3).

	Unquenched QM		Meson-Cloud
	0-5 $\hbar\omega$	0 $\hbar\omega$	Eq. (7)
$N\pi$	0.370	0.336	$4a^2/3N^2$
$\Delta\pi$	−0.027	−0.020	$-2b^2/3N^2$
$N\eta_8$, $N\eta_1$	0.007	0.0012	$4(c^2+d^2)/3N^2$
ΣK, ΛK	0.016	0.004	
Pseudoscalar	0.367	0.321	
Vector	−0.043	−0.011	
Total	0.324	0.310	

Table 5. Contribution of the sea quark spins $\Delta\Sigma_{\rm sea}$ to the spin of the proton.

	Unquenched QM		Meson-Cloud
	0-5 $\hbar\omega$	0 $\hbar\omega$	Eq. (8)
$N\pi$	−0.089	−0.084	$-a^2/3N^2$
$\Delta\pi$	0.074	0.049	$5b^2/3N^2$
$N\eta_8$, $N\eta_1$	0.006	−0.0003	$-(c^2+d^2)/3N^2$
ΣK, ΛK	0.013	0.002	
Pseudoscalar	0.005	−0.033	
Vector	0.293	0.052	
Total	0.298	0.019	

valence and sea quarks are related by $SU(3)$ flavor symmetry, allows to express the spin content of the Λ hyperon in terms of that of the proton [35, 36, 37] and gives rise to equal contributions of the quark spins $(\Delta\Sigma)_\Lambda = (\Delta\Sigma)_p$. In the unquenched quark model there is no need to make additional assumptions about the nature of the sea. Table 3 shows that the contribution of quark spins for the Λ is larger than that for the proton, $(\Delta\Sigma)_\Lambda > (\Delta\Sigma)_p$, which is a result of $SU(3)$ flavor breaking by the sea quarks.

5. Summary and conclusions

In this contribution, we presented some recent work on unquenching the quark model in which the effects of hadron loops are taken into account via a $^3P_0$ pair creation model. In particular, we studied the spin and flavor content of the proton and presented an analysis of the numerical results by means of a simple exactly solvable meson-cloud model including pion and eta loops.

The inclusion of the $q\bar{q}$ pairs leads automatically to an excess of $\bar{d}$ over $\bar{u}$ quarks, in agreement with the observed flavor asymmetry of the proton. The results for the flavor asymmetry of the proton are dominated by the $N\pi$ channel, but with important contributions from the $\Delta\pi$ channel and the off-diagonal $N\pi$-$N\eta$ terms. The contributions from orbitally excited intermediate baryons and mesons is small.

Similarly, the inclusion of hadron loops leads to a sizeable contribution of the orbital angular momentum to the spin of the proton ($\sim$ 32 %). Just as in the case of the flavor asymmetry, the contribution of orbital angular momentum to the spin of the proton is dominated by pion loops with relatively small contributions from the other channels. However, the contribution of sea quark spins to the spin of the proton ($\sim$ 30 %) is almost entirely due to excited vector mesons.

We note, that the latter contribution is absent in meson-cloud models.

Even though different models of hadron structure may show similar results for the properties of the proton, often their predictions for the other octet baryons exhibit large variations. Therefore, in order to be able to distinguish between the predictions of different models of hadron structure and to obtain a better understanding of the non-perturbative structure of QCD new experiments are needed to measure the flavor asymmetry and spin content of other octet baryons.

Acknowledgments

This work was supported in part by grants from CONACyT (78833) and PAPIIT-DGAPA (IN113711), Mexico, and INFN, Italy.

References

[1] Garvey G T and Peng J C 2001 *Prog. Part. Nucl. Phys.* **47** 203
[2] Acha A *et al* 2007 *Phys. Rev. Lett.* **98** 032301
[3] Baunack S *et al* 2009 *Phys. Rev. Lett.* **102** 151803
[4] Bass S D 2008 *The spin structure of the proton* (Singapore: World Scientific)
[5] Aznauryan I G *et al* 2008 *Phys. Rev.* C **78** 045209
Aznauryan I G *et al* 2009 *Phys. Rev.* C **80** 055203
[6] Capstick S and Roberts W 1994 *Phys. Rev.* D **49** 4570
[7] Bijker R, Iachello F and Leviatan A 1997 *Phys. Rev.* D **55** 2862
Bijker R, Iachello F and Leviatan A 2000 *Ann. Phys. (N.Y.)* **284** 89
[8] Bijker R and Santopinto E 2009 *Phys. Rev.* C **80** 065210
[9] Bijker R and Santopinto E 2010 *Phys. Rev.* C **82** 062202(R)
[10] Bijker R and Santopinto E 2010 *J. Phys.: Conf. Ser.* **239** 012009
[11] Zenczykowski P 1986 *Ann. Phys. (N.Y.)* **169** 453
[12] Geiger P and Isgur N 1990 *Phys. Rev.* D **41** 1595
[13] Geiger P and Isgur N 1993 *Phys. Rev.* D **47** 5050
[14] Geiger P and Isgur N 1997 *Phys. Rev.* D **55** 299
[15] Roberts W and Silvestre-Brac B 1992 *Few-Body Systems* **11** 171
[16] Gottfried K 1967 *Phys. Rev. Lett.* **18** 1174
[17] Amaudruz P *et al* 1991 *Phys. Rev. Lett.* **66** 2712
Arneodo M *et al* 1997 *Nucl. Phys.* B **487** 3
[18] Baldit A *et al* 1994 *Phys. Lett.* B **332** 244
[19] Towell R S *et al* 2001 *Phys. Rev.* D **64** 052002
[20] Ackerstaff K *et al* 1998 *Phys. Rev. Lett.* **81** 5519
[21] Thomas A W 1983 *Phys. Lett.* B **126** 97
Henley E M and Miller G A 1990 *Phys. Lett.* B **251** 453
[22] Eichten E J, Hinchcliffe I and Quigg C 1992 *Phys. Rev.* D **45** 2269
[23] Zhang Y J, Deng W Z and Ma B Q 2002 *Phys. Rev.* D **65** 114005
[24] Alberg M, Henley E M, Ji X and Thomas A W 1996 *Phys. Lett.* B **389** 367
Alberg M, Falter T and Henley E M 1998 *Nucl. Phys.* A **644** 93
[25] Shao L, Zhang Y J and Ma B Q 2010 *Phys. Lett.* B **686** 136
[26] Osipenko R, private communication.
[27] Ashman J *et al* 1988 *Phys. Lett.* B **206** 364
[28] Airapetian A *et al* 2007 *Phys. Rev.* D **75** 012007
[29] Alexakhin V Yu *et al* 2007 *Phys. Lett.* B **647** 8
[30] Bradamante F 2008 *Prog. Part. Nucl. Phys.* **61** 229
[31] Sehgal L M 1974 *Phys. Rev.* D **10** 1663
[32] Ratcliffe P G 1987 *Phys. Lett.* B **192** 180
[33] Myhrer F and Thomas A W 2008 *Phys. Lett.* B **663** 302
[34] Garvey G T 2010 *Preprint* arXiv:1001.4547 [nucl-th]
[35] Jaffe R L 1996 *Phys. Rev.* D **54** R6581
[36] Boros C and Zuo-tang L 1998 *Phys. Rev.* D **57** 4491
[37] Ashery D and Lipkin H J 1999 *Phys. Lett.* B **469** 263

Journal of Physics: Conference Series **322** (2011) 012015 doi:10.1088/1742-6596/322/1/012015

Quark-model study of the hadron structure and the hadron-hadron interaction

A Valcarce

Departamento de Física Fundamental, Universidad de Salamanca, E-37008 Salamanca, Spain

J Vijande

Departamento de Física Atómica, Molecular y Nuclear, Universidad de Valencia (UV) and IFIC (UV-CSIC), Valencia, Spain.

H Garcilazo

Escuela Superior de Física y Matemáticas, Instituto Politécnico Nacional, Edificio 9, 07738 México D.F., Mexico

T F Caramés

Departamento de Física Fundamental, Universidad de Salamanca, E-37008 Salamanca, Spain

Abstract. Recent results of hadron spectroscopy and hadron-hadron interaction within a quark model framework are reviewed. Higher order Fock space components are considered based on new experimental data on low-energy hadron phenomenology. The purpose of this study is to obtain a coherent description of the low-energy hadron phenomenology to constrain QCD phenomenological models and try to learn about low-energy realizations of the theory.

1. Introduction

We present an overview about the status of the old dream of getting a coherent description of the low-energy hadron phenomenology in a quark model scheme [1]. We make emphasis on two different aspects. Firstly, we will address the study of meson and baryon spectroscopy in an enlarged Hilbert space considering four and five quark components. This seems nowadays unavoidable to understand the experimental data and it builds the bridge towards the description in terms of hadronic degrees of freedom. Secondly, the same scheme used to describe the hadron spectroscopy should be valid for describing the low-energy hadron-hadron scattering. We review our recent efforts to account for the strangeness -1 and -2 two- and three-baryon systems within the same scheme used for hadron spectroscopy and that has also demonstrated its validity for the nucleon-nucleon sector.

In particular we will address three different problems. In the first section we will discuss the charmonium spectroscopy above the threshold for the production of D mesons. The second section is devoted to study the effect of selected S-wave meson-baryon channels to match poor baryon mass predictions from quark models with data. Finally, the third section addresses the problem of the study of two- and three-baryon systems with strangeness -1 and -2 by

Journal of Physics: Conference Series **322** (2011) 012015 doi:10.1088/1742-6596/322/1/012015

means of a baryon-baryon interaction derived from the quark model used to study the hadron spectroscopy.

2. Heavy mesons: Charmonium

Charmonium has been used as the test bed to demonstrate the color Fermi-Breit structure of quark atoms obeying the same principles as ordinary atoms [2]. Its nonrelativistic character ($v/c \approx 0.2 - 0.3$) gave rise to an amazing agreement between experiment and simple quark potential model predictions as $c\bar{c}$ states [3]. Close to the threshold for the production of charmed mesons quark models required of an improved interaction [4].

Since 2003, we have witnessed a growth of puzzling new charmonium mesons, like the well-established $X(3872)$. These new states do not fit, in general, the simple predictions of the quark-antiquark schemes and, moreover, they overpopulate the expected number of states in (simple) two-body theories. This situation is not uncommon in particle physics. For example, in the light scalar-isoscalar meson sector hadronic molecules offer a reasonable explanation of the experimental data [5]. Also, the study of the NN system above the pion production threshold required new degrees of freedom to be incorporated in the theory, either as pions or as excited states of the nucleon, i.e., the Δ [6]. This discussion suggests that charmonium spectroscopy could be rather simple below the threshold production of charmed mesons but much more complex above it. In particular, the coupling to the closest $(c\bar{c})(n\bar{n})$ system, referred to as *unquenching the naïve quark model* [7], could be an important spectroscopic ingredient. Besides, hidden-charm four-quark states could explain the overpopulation of quark-antiquark theoretical states.

In an attempt to disentangle the role played by multiquark configurations in the charmonium spectroscopy we have solved the Lippmann-Schwinger equation for the scattering of two D mesons looking for attractive channels that may contain a meson-meson molecule [8]. In order to account for all basis states we allow for the coupling to charmonium-light two-meson systems. We have consistently used the same interacting Hamiltonian to study the two- and four-quark systems to guarantee that thresholds and possible bound states are eigenstates of the same Hamiltonian. Our predictions robustly show that no deeply bound states can be expected for this system. Only a few channels, see Table 1, can be expected to present observable resonances or slightly bound states. Among them, we have found that the $D\overline{D}^*$ system must show a bound state slightly below threshold with quantum numbers $J^{PC}(I) = 1^{++}(0)$, that could correspond to the widely discussed $X(3872)$.

Out of the systems made of a particle and its corresponding antiparticle, $D\overline{D}$ and $D^*\overline{D}^*$, the $J^{PC}(I) = 0^{++}(0)$ is attractive. It would be the only candidate to accommodate a wide resonance for the $D\overline{D}$ system. For the $D^*\overline{D}^*$ the attraction is stronger and structures may be observed close and above the charmed meson production threshold. We have shown that the $J^{PC}(I) = 2^{++}(0,1)$ $D^*\overline{D}^*$ channels are attractive due to the coupling to the $J/\Psi\omega$ and $J/\Psi\rho$ channels, respectively. The charged states reported in Ref. [9] could correspond to the $D^*\overline{D}^*$ $J^{PC}(I) = 2^{++}(1)$ we have predicted [10]. Its confirmation would represent a unique tool in discriminating among different theoretical models.

Table 1. Attractive channels for the two $D-$meson systems.

System	$D\overline{D}$	$D\overline{D}^*$	$D^*\overline{D}^*$		
$J^{PC}(I)$	$0^{++}(0)$	$1^{++}(0)$	$0^{++}(0)$	$2^{++}(0)$	$2^{++}(1)$

3. Light baryons

As we have seen above, there is nowadays compelling evidence of meson resonances containing more than quark-antiquark ($q\bar{q}$) valence components. One can guess a similar situation for some baryon resonances that may contain more than three-quark ($3q$) valence components. In these cases valence quark models based on $3q$ or $q\bar{q}$ components that describe correctly the bulk of spectral data fail systematically to reproduce their masses. Actually we can interpret the systematic failure in the description of a given resonance by valence quark models as an indication of its anomalous nature in the sense of containing more than the valence components. Here we use this interpretation as a criteria to identify possible anomalies in the Light-quark Baryon Spectrum (LqBS).

A comparative analysis of Constituent Quark Models (CQM) predictions for the LqBS has been carried out in Ref. [12]. This analysis makes clear the presence of anomalies for which $3q$ predicted masses are significantly higher than their Particle Data Group (PDG) averages [13], see Fig. 1. Moreover all the anomalies have in common the existence of $S-$ wave meson-baryon thresholds with the same quantum numbers than the anomalies (also drawn in Fig. 1) close above the PDG average masses. This suggests a possible contribution of these meson-baryon ($4q1\bar{q}$) components to the anomalies to give a correct account of their masses. To make a rough estimate of this contribution we shall consider the coupling of the free meson-baryon (mb) channels to the corresponding $3q$ states. To make effective such a coupling we parametrize the transition matrix element, $< mb|H|3q >\equiv a$, and diagonalize the hamiltonian matrix by assuming $< 3q|H|3q >= M_{3q}$ being M_{3q} the predicted $3q$ mass and $< mb|H|mb >\sim M_m + M_b$ being $M_m + M_b$ the mass of the meson-baryon threshold. The results from the diagonalization corresponding to the lowest energy states are shown in Fig. 2. Despite the shortcomings of our toy model calculation the quantitative agreement with data is spectacular (there is only one universal parameter, a = 85 MeV, assumed for simplicity to be the same in all cases). Qualitatively our toy model implies that the anomalies correspond mostly to meson-baryon

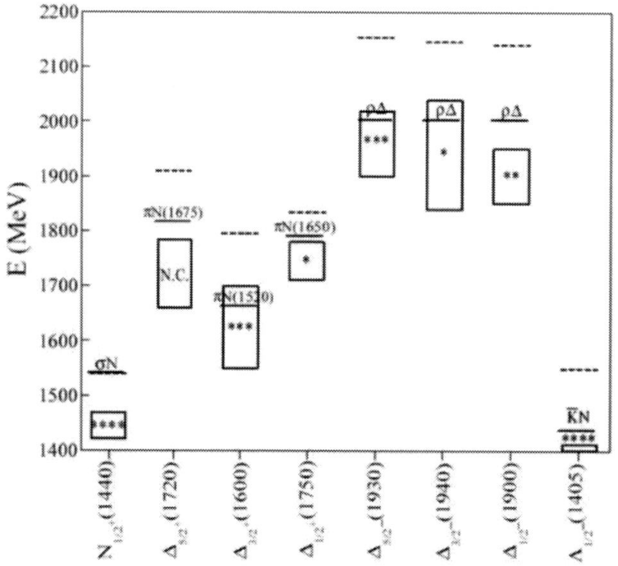

Figure 1. Mass predictions for the anomalies from Ref. [11] (dashed lines) as compared to the experimental mass intervals (boxes). N.C. means non-cataloged resonance. Meson-baryon thresholds are indicated by solid lines.

Journal of Physics: Conference Series **322** (2011) 012015 doi:10.1088/1742-6596/322/1/012015

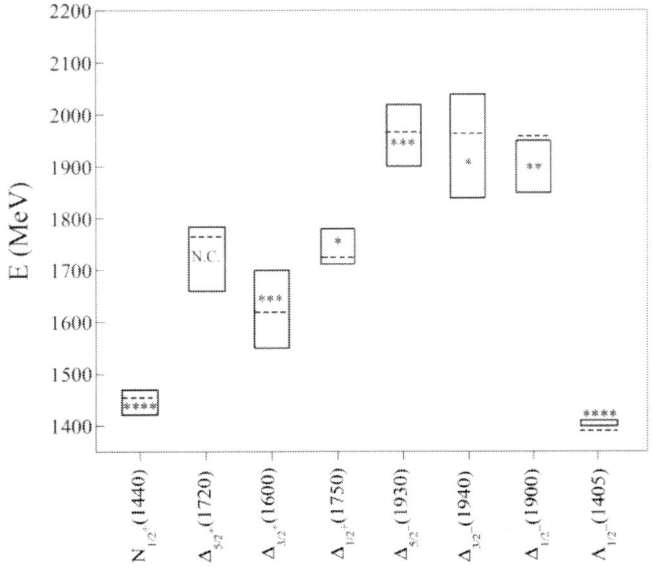

Figure 2. Predicted masses for the anomalies (dashed lines) as compared to the experimental mass intervals (boxes). N.C. means non-cataloged resonance.

states but with a non-negligible three-quark probability that makes them stable against decay into $m+b$. This interpretation could need some refinement when going to a more complete model where the meson-baryon interaction were also considered ($< mb|H|mb >\sim M_m + M_b$). Then some anomalies might appear as simply meson-baryon bound states. Keeping this in mind we may conclude that the coupling to selected mb channels acts as an effective healing mechanism to match deficient CQM predictions with data. Furthermore the implementation of these selected channels in the analyses of data might add certainty to the existence of the anomalies and at the same time help to reconcile competing and sometimes not compatible partial wave analyses.

4. Strangeness -1 and -2 two and three-baryon systems

In Ref. [14] we constructed different families of $S = -1$ interacting potentials, by introducing small variations of the mass of the effective scalar exchange potentials, that allow us to study the dependence of the results on the strength of the spin-singlet and spin-triplet hyperon-nucleon interactions. These potentials are characterized by the ΛN scattering lengths $a_{i,s}$ and they reproduce the cross sections near threshold of the five hyperon-nucleon processes for which data are available. Let us analyze the three-body systems with strangeness -1.

4.1. Strangeness -1 three-baryon systems: ΛNN and ΣNN

The channels (I, J) = (0,1/2) and (0,3/2) are the most attractive ones of the ΛNN system. In particular, the channel (0,1/2) has the only bound state of this system, the hypertriton. We give in Table 2 the results of the models constructed in Ref. [14] for the two Λd scattering lengths and the hypertriton binding energy. We compare with the results, in parentheses, obtained in Ref. [14] including only the three-body S wave configurations. As a consequence of considering the D waves, the hypertriton binding energy increases by about 50–60 keV [15], while the $A_{0,1/2}$ scattering length decreases by about 3–5 fm. The largest changes occur in the $A_{0,3/2}$ scattering length where both positive and negative values appeared which means, in the case of the negative values, that a bound state is generated in the $(I, J) = (0, 3/2)$ channel. Since this channel depends mainly on the spin-triplet hyperon-nucleon interaction and experimentally

Table 2. Λd scattering lengths, $A_{0,3/2}$ and $A_{0,1/2}$ (in fm), and hypertriton binding energy, $B_{0,1/2}$ (in MeV), for several hyperon-nucleon interactions characterized by ΛN scattering lengths $a_{1/2,0}$ and $a_{1/2,1}$ (in fm).

$a_{1/2,0}$	$a_{1/2,1}$	$A_{0,3/2}$	$A_{0,1/2}$	$B_{0,1/2}$
2.48	1.41	31.9 (66.3)	−16.0 (−20.0)	0.129 (0.089)
2.48	1.65	−72.8 (198.2)	−13.8 (−17.2)	0.178 (0.124)
2.48	1.79	−28.5 (−62.7)	−12.9 (−16.0)	0.207 (0.145)
2.31	1.65	−76.0 (198.2)	−17.1 (−22.4)	0.113 (0.070)
2.74	1.65	−72.1 (198.2)	−12.0 (−14.4)	0.244 (0.182)

there is no evidence whatsoever for the existence of a $(I,J) = (0,3/2)$ bound state one can use the results of this channel to set limits on the value of the hyperon-nucleon spin-triplet scattering length $a_{1/2,1}$. As one can guess from Table 2, the three-body channel $(I,J) = (0,3/2)$ becomes bound if $a_{1/2,1} > 1.58$ fm (see Ref. [16]. Moreover, we found in Ref. [14] that the fit of the hyperon-nucleon cross sections is worsened for those cases where the spin-triplet ΛN scattering length is smaller than 1.41 fm, so that we conclude that $1.41 \leq a_{1/2,1} \leq 1.58$ fm. To set some limits to the hyperon-nucleon spin-singlet scattering length, we have calculated in Table 3 the hypertriton binding energy using for the hyperon-nucleon spin-triplet scattering length the allowed values $1.41 \leq a_{1/2,1} \leq 1.58$ fm and for the spin-singlet scattering length $2.33 \leq a_{1/2,0} \leq 2.48$ fm which leads to a hypertriton binding energy within the experimental error bars $B_{0,1/2} = 0.13 \pm 0.05$ MeV.

We show in Table 4 the Σd scattering lengths $A'_{1,3/2}$ and $A'_{1,1/2}$. The Σd scattering lengths are complex since the inelastic ΛNN channels are always open. The scattering length $A'_{1,3/2}$ depends mainly on the spin-triplet hyperon-nucleon channels and both its real and imaginary parts increase when the spin-triplet hyperon-nucleon scattering length increases. The effect of the three-body D waves is to lower the real part by about 20 % and the imaginary part by about 10 %. The scattering length $A'_{1,1/2}$ shows large variations between the results with and without three-body D waves, due to the presence of a pole very near threshold. The position of this pole in the complex plane, given in the last column of Table 4, changes very little with the model used and it lies at around $2.8 - i\,2.1$ MeV.

4.2. Strangeness -2 two-baryon systems: $\Lambda\Lambda$, $\Sigma\Sigma$ and $N\Xi$

In this section, we derive the strangeness −2 baryon-baryon interactions: $\Lambda\Lambda$, $\Lambda\Sigma$, $\Sigma\Sigma$ and $N\Xi$. We use these two-body interactions to calculate two-body elastic and inelastic scattering cross sections [17] and we compare to experimental data and other theoretical models. We also

Table 3. Hypertriton binding energy (in MeV) for several hyperon-nucleon interactions characterized by ΛN scattering lengths $a_{1/2,0}$ and $a_{1/2,1}$ (in fm) which are within the experimental error bars $B_{0,1/2} = 0.130 \pm 0.050$ MeV.

	$a_{1/2,1} = 1.41$	$a_{1/2,1} = 1.46$	$a_{1/2,1} = 1.52$	$a_{1/2,1} = 1.58$
$a_{1/2,0} = 2.33$	0.080	0.087	0.096	0.106
$a_{1/2,0} = 2.39$	0.094	0.102	0.112	0.122
$a_{1/2,0} = 2.48$	0.129	0.140	0.152	0.164

Table 4. Σd scattering lengths, $A'_{1,3/2}$ and $A'_{1,1/2}$ (in fm), and position of the quasibound state $B'_{1,1/2}$ (in MeV) for the hyperon-nucleon interactions of Table 2.

$a_{1/2,0}$	$a_{1/2,1}$	$A'_{1,3/2}$	$A'_{1,1/2}$	$B'_{1,1/2}$
2.48	1.41	$0.14+i\,0.24$ $(0.20+i\,0.26)$	$19.82+i\,16.94$ $(19.28+i\,25.37)$	$2.92-i\,2.17$
2.48	1.65	$0.28+i\,0.27$ $(0.36+i\,0.29)$	$12.08+i\,38.98$ $(-1.55+i\,42.31)$	$2.84-i\,2.14$
2.48	1.79	$0.36+i\,0.29$ $(0.44+i\,0.31)$	$-8.00+i\,42.58$ $(-17.33+i\,35.01)$	$2.79-i\,2.10$
2.31	1.65	$0.28+i\,0.27$ $(0.36+i\,0.29)$	$19.01+i\,23.21$ $(14.95+i\,31.61)$	$2.88-i\,2.14$
2.74	1.65	$0.28+i\,0.27$ $(0.36+i\,0.29)$	$-26.01+i\,17.95$ $(-23.29+i\,13.32)$	$2.73-i\,2.09$

calculate the two-body scattering lengths of the different spin-isospin channels to compare with other theoretical models. We have calculated [17] the $\Xi^- p$ elastic cross section compared to the in-medium experimental $\Xi^- p$ cross section around $p^{\Xi}_{\rm lab} = 550$ MeV/c, where $\sigma_{\Xi^- p} = 30 \pm 6.7\ ^{+3.7}_{-3.6}$ mb [18]. Another analysis using the eikonal approximation gives $\sigma_{\Xi^- p} = 20.9 \pm 4.5\ ^{+2.5}_{-2.4}$ mb [19]. A more recent experimental analysis [20] for the low energy $\Xi^- p$ elastic and $\Xi^- p \to \Lambda\Lambda$ total cross sections in the range 0.2 GeV/c to 0.8 GeV/c shows that the former is less than 24 mb at 90% confidence level and the latter of the order of several mb, respectively. We also evaluated in Ref. [17] the inelastic $\Xi^- p \to \Lambda\Lambda$ cross section. It has been recently estimated at a laboratory momentum of $p^{\Xi}_{\rm lab} = 500$ MeV/c, see Ref. [20], assuming a quasifree scattering process for the reaction ^{12}C$(\Xi^-, \Lambda\Lambda)X$ obtaining a total cross section $\sigma(\Xi^- p \to \Lambda\Lambda) = 4.3^{+6.3}_{-2.7}$ mb. The upper limit of the cross section was derived as 12 mb at 90% confidence level. We would like to emphasize the agreement of our results with the $\Xi^- p \to \Lambda\Lambda$ conversion cross section. This reaction is of particular importance in assessing the stability of Ξ^- quasi-particle states in nuclei.

The scattering lengths for the different spin-isospin channels are given in Table 5. As mentioned above, since the observation of the Nagara event [24] it is generally accepted that the $\Lambda\Lambda$ interaction is only moderately attractive. Our result for the $^1S_0$ $\Lambda\Lambda$ scattering length is compatible with such event. A rough estimate of $B_{\Lambda\Lambda}(B_{\Lambda\Lambda} \approx (\hbar c)^2/2\mu_{\Lambda\Lambda}a^{\Lambda\Lambda}_{^1S_0})$ drives a value of 5.41 MeV, below the upper limit extracted from the Nagara event, $B_{\Lambda\Lambda} = 7.25 \pm 0.19$ MeV. Moreover, although from the $\Lambda\Lambda$ scattering length alone one cannot draw any conclusion on the magnitude of the two-Λ separation energy, recent estimates [25] have reproduced the two-Λ separation energy, defined as $\Delta B_{\Lambda\Lambda} = B_{\Lambda\Lambda}(^{6}_{\Lambda\Lambda}\mathrm{He}) - 2B_{\Lambda}(^{5}_{\Lambda}\mathrm{He})$, with scattering lengths of -1.32 fm.

Table 5. Two-body singlet and triplet scattering lengths, in fm, for different models as compared to our results. The results between squared brackets indicate the lower and upper limit for different parametrizations used in that reference.

Model	Ref. [21]	Ref. [22]	Ref. [23]	Ref. [15]	Ours [17]
$a^{\Lambda\Lambda}_{^1S_0}$	$[-0.27, -0.35]$	$[-1.555, -3.804]$	$[-1.52, -1.67]$	-0.821	-2.54
$a^{\Xi^0 p}_{^1S_0}$	$[0.40, 0.46]$	$[0.144, 0.491]$	$[0.13, 0.21]$	0.324	-3.32
$a^{\Xi^0 p}_{^3S_1}$	$[-0.030, 0.050]$	–	$[0.0, 0.03]$	-0.207	18.69
$a^{\Sigma^+\Sigma^+}_{^1S_0}$	$[6.98, 10.32]$	–	$[-6.23, -9.27]$	-85.3	0.523

5. Summary

To summarize, we have reviewed some recent results of hadron spectroscopy and hadron-hadron interaction in terms of the constituent quark model. The main goal of our presentation was to highlight the complementarity of a simultaneous study of both problems in order to constrain low-energy realizations of the underlying theory, QCD. We have seen how the enlargement of the Hilbert space when increasing energy, that was seen to be necessary to describe the nucleon-nucleon phenomenology above the pion production threshold, seems to be necessary to understand current problems of hadron spectroscopy. We have also tried to emphasize that spectroscopy and interaction can and must be understood within the same scheme when dealing with quark models. Any other alternative becomes irrelevant from the point of view of learning about properties of QCD.

In our analysis we have shown the first systematic study of four-quark hidden-charm states as meson-meson molecules. Our predictions robustly show that no deeply bound states can be expected for charmonium. Only a few channels can be expected to present observable resonances or slightly bound states. Among them, we have found that the $D\overline{D}^*$ system must show a bound state slightly below the threshold for charmed mesons production with quantum numbers $J^{PC}(I) = 1^{++}(0)$, that could correspond to the widely discussed $X(3872)$. We have also shown that the $J^{PC}(I) = 2^{++}(1)$ $D^*\overline{D}^*$ channel is attractive and may then represent a charge state contributing to charmonium spectroscopy.

We have seen how the $4q1\overline{q}$ components, in the form of S wave meson-baryon channels which we identify, play an essential role in the description of the anomalies, say baryon resonances very significantly overpredicted by three-quark models based on two-body interactions. Considering a simplified description of the anomalies as systems composed of a free meson-baryon channel interacting with a three-quark confined component we have shown they could correspond mostly to meson-baryon states but with a non-negligible $3q$ state probability which makes their masses to be below the meson-baryon threshold.

We also presented results for three-baryon systems with strangeness -1 using the hyperon-nucleon and nucleon-nucleon interactions derived from a constituent quark model with full inclusion of the $\Lambda \leftrightarrow \Sigma$ conversion. In the case of the ΛNN system, our calculation permits to constrain the ΛN scattering lengths to: $1.41 \leq a_{1/2,1} \leq 1.58$ fm and $2.33 \leq a_{1/2,0} \leq 2.48$ fm. In the case of the ΣNN system there exists a narrow quasibound state near threshold in the $(I, J) = (1, 1/2)$ channel.

We have presented results for the doubly strange two-baryon interactions. Our predictions are consistent with the recently obtained doubly strange elastic and inelastic scattering cross sections. In particular, our results are compatible with the $\Xi^- p \to \Lambda\Lambda$ conversion cross section, important in assessing the stability of Ξ^- quasi-particle states in nuclei.

Acknowledgments

This work has been partially funded by the Spanish Ministerio de Educación y Ciencia and EU FEDER under Contracts No. FPA2007-65748 and FPA2010-21750, and by the Spanish Consolider-Ingenio 2010 Program CPAN (CSD2007-00042).

References

[1] Valcarce A, Garcilazo H, Fernández F and González P 2005 *Rep. Prog. Phys.* **68** 965
[2] Isgur N and Karl G 1983 *Physics Today* November 36
[3] Eichten E, Gottfried K, Kinoshita T, Lane K D and Yan T M 1980 *Phys. Rev. D* **21** 203
[4] Isgur N 1999 *Phys. Rev. D* **60** 054013
[5] Jaffe R L 1977 it Phys. Rev. D **15** 267
[6] Pöpping H, Sauer P U and Zhang X Z 1987 *Nucl. Phys. A* **474** 557
Valcarce A, Fernández F, Garcilazo H, Peña M T and Sauer P U 1994 *Phys. Rev. C* **49** 1799
[7] Close F E 2007 *Preprint* arXiv:0706.2709.

Journal of Physics: Conference Series **322** (2011) 012015 doi:10.1088/1742-6596/322/1/012015

[8] Fernández-Caramés T, Valcarce A and Vijande J 2009 *Phys. Rev. Lett.* **103** 222001
[9] Mizuk R *et al* 2008 *Phys. Rev. D* **78** 072004
[10] Caramés T, Valcarce A and J. Vijande J *Phys. Rev. D* **82** 054032
[11] Capstick S and Isgur N 1986 *Phys. Rev. D* **34** 2809
[12] González P, Vijande J and Valcarce A 2008 *Phys. Rev. C* **77** 065213
[13] Amsler C *et al* 2008 *Phys. Lett. B* **667** 1
[14] Garcilazo H, Fernández-Caramés T, and Valcarce A 2007 *Phys. Rev. C* **75** 034002
[15] Fujiwara Y, Suzuki Y and Nakamoto C 2007 *Prog. Part. Nucl. Phys.* **58** 439
[16] Garcilazo H, Valcarce A and Fernández-Caramés T 2007 *Phys. Rev. C* **76** 034001
[17] Valcarce A, Garcilazo H and Caramés T F 2010 *Phys. Lett. B* **693** 305
[18] Tamagawa T *et al* 2001 *Nucl. Phys. A* **691** 234c
[19] Yamamoto Y, Tamagawa T, Fukuda T and Motoba T 2001 *Prog. Theor. Phys.* **106** 363
[20] Ahn J K *et al* 2006 *Phys. Lett. B* **633** 214
[21] Stoks V G J and Rijken Th A 1999 *Phys. Rev. C* **59** 3009
[22] Rijken Th A and Yamamoto Y 2006 *Preprint* nucl-th/0608074
[23] Polinder H, Haidenbauer J, and Meissner U G 2007 *Phys. Lett. B* **653** 29
[24] Takahashi H *et al* 2001 *Phys. Rev. Lett.* **87** 212502
[25] Rijken Th A and Yamamoto Y 2006 *Phys. Rev. C* **73** 044008

Journal of Physics: Conference Series **322** (2011) 012016 doi:10.1088/1742-6596/322/1/012016

QCD at low energy: a many-body approach

T Yépez-Martínez[1], P O Hess[1], S Lerma[2], A Szczepaniak[3] and O Civitarese[4]

[1] Instituto de Ciencias Nucleares, Universidad Nacional Autónoma de México, AP 70-543, 04510 Mexico DF, Mexico
[2] Departamento de Fśica, Facultad de Física e Inteligencia Artificial, Universidad Veracruzana, Lomas del Estadio s/n, C.P. 91000, Xalapa, Veracruz, México
[3] Department of Physics and Nuclear Theory Center, Indiana University Bloomington, Indiana, 47405-4202, USA
[4] Departamento de Física, Universidad Nacional La Plata, C.C. 67 (1900), La Plata, Argentina

E-mail: `tochtli.yepez@nucleares.unam.mx`, `hess@nucleares.unam.mx`, `lerma@uv.mx`, `aszczepa@indiana.edu`, `osvaldo.civitare@fisica.unlp.edu.ar`

Abstract. A review is given on recent results in the treatment of an arbitrary number of orbital levels in low energy QCD. For the pure quark part, analytic results for the dominant part of the Hamiltonian are presented. Possible extensions, including dynamic gluons, are discussed.

1. Introduction

QCD is considered as the correct theory for strong interactions [1, 2]. Unfortunately, at low energy the coupling constant is very large and perturbative calculations are impossible. Up to now, only lattice-QCD [3] is able to derive non-perturbative results from first principles. Lattice calculations require an enormous numerical effort and results, like the level ordering of the spectrum, are not well understood. As an alternative, phenomenological models were proposed [4, 5, 6] (just to name a few) which have the advantage of being easy to apply and transparency in their physical interpretation, but with the disadvantage of involving several parameters. In [7, 8] we presented a method on how to obtain analytical and semi-analytical (involving a simple numerical application) solution.

In this contribution we will shortly resume the results of [7, 8], concerning the semi-analytical solution including an arbitrary number of orbital levels for quarks and anti-quarks. Gluons are considered to be static. After that, we will discuss on possible methods on how include gluons dynamically.

2. The Hamiltonian and the basic Ingredients

The QCD Hamiltonian at low energy is given by [9]

$$\begin{aligned}
\boldsymbol{H} &= \boldsymbol{H}_0 + \boldsymbol{H}_{\rm int} \; , \\
\boldsymbol{H}_0 &= \int d^3x \boldsymbol{\psi}^\dagger \left(-i\boldsymbol{\alpha}\cdot\boldsymbol{\nabla} + \boldsymbol{\beta} m_0 \right) \boldsymbol{\psi} \\
&+ \frac{1}{2}\int d^3x \boldsymbol{\Pi}^2 - \frac{1}{2}\int d^3x \boldsymbol{A}\cdot\boldsymbol{\nabla}\boldsymbol{A} + V_0 \int d^2x d^3y \boldsymbol{\rho}_a(\vec{x})\boldsymbol{\rho}^a(\vec{y}) \; , \\
\boldsymbol{H}_{\rm int} &= \int d^2x d^3y \boldsymbol{\rho}_a(\vec{x}) \left(\frac{1}{2} K^{(0)}(\vec{x}-\vec{y}) - V_0 \right) \boldsymbol{\rho}^a(\vec{y}) \\
&+ \frac{1}{2}\int d^3x \left[\boldsymbol{B}^2 + \boldsymbol{A}\cdot\boldsymbol{\nabla}^2\boldsymbol{A} \right] - g\int d^x \boldsymbol{\psi}^\dagger \boldsymbol{\alpha}\cdot\boldsymbol{A}\boldsymbol{\psi} + \boldsymbol{V}_A + \boldsymbol{V}_B \\
&+ \frac{1}{2}\int d^3x d^3y \boldsymbol{\rho}_a(\vec{x}) \left[K^{ab}(\vec{x}-\vec{y};\boldsymbol{A}) - \delta^{ab}K^{(0)}(\vec{x}-\vec{y}) \right] \boldsymbol{\rho}_b(\vec{y}) \; , \qquad (1)
\end{aligned}$$

where we divided the Hamiltonian in the dominant $\boldsymbol{H}_0$ and a residual part $\boldsymbol{H}_{int}$. The $\boldsymbol{H}_0$ part contains the kinetic and mass terms of the quark fields ψ, the proposed kinetic gluon term and the color charge density interaction which is a remanent of the Coulomb-QCD interaction. The terms related to the Coulomb interaction are called such, because in QED they reduce to the Coulomb potential [1]. The $\boldsymbol{V}_A$ and $\boldsymbol{V}_B$ in the interacting part are involved functions depicted in [9], which contain terms of the expansion of the inverse of the Fadeev-Popov term $(\boldsymbol{\nabla}\cdot\boldsymbol{D})$ in terms of the gluon fields, since $\boldsymbol{D}$ is the covariant derivative and contains the gluon field. In contrast to [9], the gluon potential $K^{(0)}(\mid \vec{x}-\vec{y} \mid)$ is approximated by a constant and the difference, $(\frac{1}{2}K^{(0)}(\vec{x}-\vec{y}) - V_0)$, is moved to the residual part of the Hamiltonian. This approximation holds in a certain low energy region (about 0 to 1.5 GeV), to which we will restrict. The other terms in the last two lines of eq. (1) corresponds to the cromo-magnetic and the quark-gluon terms, and the residual Coulomb-QCD interaction respectively. This residual Coulomb-QCD interaction corresponds to the difference between the real Coulomb-QCD kernel $K^{ab}(\vec{x}-\vec{y};\boldsymbol{A})$ and its vacuum expectation value $(K^{(0)}(\vec{x}-\vec{y})\delta^{ab})$.

In [7, 8] only the dominant part, $\boldsymbol{H}_0$, of the Hamiltonian was considered. The K functions are kernels and some depend also on the gluon field $\boldsymbol{A}$. These functions can be expanded in the gluons fields and as shown in [10] it suffices to expand them up to third order in $\boldsymbol{A}$. The main reason is that at low energies the gluon fields represent effective gluons with a mass between 600 and 800 MeV, i.e., more gluons imply higher energies.

As basic ingredients we i) use $SU_C(3)$ for the color part, ii) $SU_F(2)$ for the flavor part and iii) static gluons only. The calculations are performed within the Coulomb gauge [1], which has the advantage of a reduced Hilbert space (no ghost contributions) but the price to pay is a complicated Hamiltonian. iv) The calculations are performed in a finite volume, of the size of a hadron. This approximation has later to be relaxed in order to study greater volumes. v) We use the Dirac picture [2], which allows easier access for many-body techniques coming from nuclear physics [11].

The degrees of freedom are described by the fermion creation operator

$$\boldsymbol{b}^\dagger_{\alpha(N,l\frac{1}{2})j\lambda,cf} = \sum_{m\sigma} (lm,\frac{1}{2}\sigma \mid j\lambda)\, \boldsymbol{b}^\dagger_{\alpha Nlm,\sigma cf} \; , \qquad (2)$$

represented here in their coupled and decoupled form in spin. The first index $\alpha = \pm\frac{1}{2}$ refer to states in the upper (+ sign) and lower (− sign) orbital level. N refers to the number of oscillation quanta of the orbital level (i.e., we use the three dimensional harmonic oscillator as the basis), l is the orbital angular momentum, j is the total spin with projection λ and c, f

Journal of Physics: Conference Series **322** (2011) 012016 doi:10.1088/1742-6596/322/1/012016

are the components of color and flavor respectively. Under raising and lowering indices, these operators satisfy [12, 13]

$$\boldsymbol{b}^{\dagger\,\alpha(N,l,\frac{1}{2})j\lambda,cf} \equiv (-1)^{\frac{1}{2}-\alpha}(-1)^{j-\lambda}(-1)^{\chi_C}(-1)^{\frac{1}{2}-f}\boldsymbol{b}^{\dagger}_{-\alpha(N,l,\frac{1}{2})j-\lambda,\bar{c}-f}\,, \tag{3}$$

where $(-1)^{\chi_C}$ is a short hand notation for the phase of color-hypercharge Y_C, color-isospin I_C and projection of the color-isospin T_{Z_C} [12].

3. Analytic and semi-analytic solutions for $\boldsymbol{H}_0$

Expressed in terms of the fermion creation and annihilation operators, restricting to the pure quark part, the potential and the kinetic energy of $\boldsymbol{H}_0$ acquire respectively the form

$$\begin{aligned}\boldsymbol{V} &= 2V_0\boldsymbol{\mathcal{C}}_2(SU(3))\,,\\ \boldsymbol{K} &= \widetilde{\boldsymbol{K}}_+ + \widetilde{\boldsymbol{K}}_- = \sum_{Nj}\left(\widetilde{\boldsymbol{K}}_+^{jN} + \widetilde{\boldsymbol{K}}_-^{jN}\right)\,,\end{aligned} \tag{4}$$

where $\boldsymbol{\mathcal{C}}_2(SU(3))$ is the second order Casimir operator of $SU_C(3)$, with eigenvalues $(\lambda_C^2 + \lambda_C\mu_C + \mu_C^2 + 3\lambda_C + 3\mu_C)$ and

$$\begin{aligned}\widetilde{\boldsymbol{K}}_+ &= \sum_{N'Nj\lambda cf}\left|k^j_{NN'}\right|\Big[\boldsymbol{b}^\dagger_{\frac{1}{2}(N,j+\frac{1}{2},\frac{1}{2})j\lambda cf}\boldsymbol{b}^{-\frac{1}{2}(N',j-\frac{1}{2},\frac{1}{2})j\lambda cf}\\ &\quad+\boldsymbol{b}^\dagger_{\frac{1}{2}(N',j-\frac{1}{2},\frac{1}{2})j\lambda cf}\boldsymbol{b}^{-\frac{1}{2}(N,j+\frac{1}{2},\frac{1}{2})j\lambda cf}\Big]\,,\\ \widetilde{\boldsymbol{K}}_- &= \sum_{N'Nj\lambda cf}\left|k^j_{NN'}\right|\Big[\boldsymbol{b}^\dagger_{-\frac{1}{2}(N,j+\frac{1}{2},\frac{1}{2})j\lambda cf}\boldsymbol{b}^{\frac{1}{2}(N',j-\frac{1}{2},\frac{1}{2})j\lambda cf}\\ &\quad+\boldsymbol{b}^\dagger_{-\frac{1}{2}(N',j-\frac{1}{2},\frac{1}{2})j\lambda cf}\boldsymbol{b}^{\frac{1}{2}(N,j+\frac{1}{2},\frac{1}{2})j\lambda cf}\Big]\,.\end{aligned} \tag{5}$$

The factors $k^j_{NN'}$ can be retrieved from [8].

The terms in (5) connect orbital levels at negative energy to the ones at positive energy, changing the orbital angular momentum. In Fig. 1 the situation is illustrated, indicating only the positive energy levels, in representation also for the negative energy levels. The terms of the kinetic energy connect only states with *the same* total spin j. As a consequence it suffices to restrict to an arbitrary but fixed j. A solution obtained for an arbitrary but fixed j implies many solutions for arbitrary j.

Restricting to two or three levels (positive and negative energy) we were able to obtain analytic results, rewriting the Hamiltonian in such a way that it exhibits an $SU(2)$ structure of pseudo-spin. This could not be repeated for more than three orbital levels.

For more than three levels, one has to apply two successive unitary transformations. The first one mixes, for a given j, only states with the same orbital spin, leading to a partial diagonalization. The matrix elements of this transformation are denoted by α^j_{Nk}, for orbital spin $l = j + \frac{1}{2}$, and β^j_{Nk}, for orbital spin $l = j - \frac{1}{2}$. The index N denotes the main oscillator quantum number and k the new orbital index. The new quark creation and annihilation operators are then a combination of all quark creation and annihilation operators of the same orbital spin, l, but different N. The second unitary transformation is a BCS transformation which diagonalizes the orbital index. For more details, please consult [8].

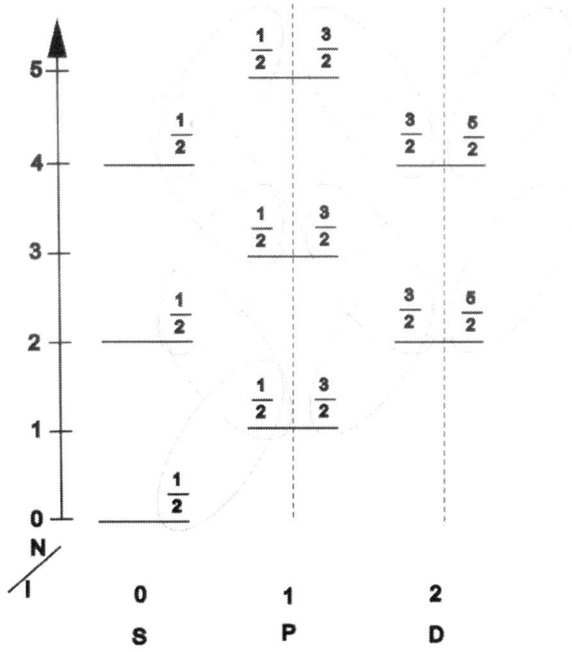

Figure 1. Illustration of the effect of the Hamiltonian on the basis states. The operators, which appear in the kinetic energy, can be divided into columns of operators, related to a given spin j, which commute with those of different columns. Only the positive energy states are plotted. The ellipses indicate which states are connected through the interaction terms appearing in the kinetic energy. Columns of different j are separated by a vertical dashed line.

For the Hamiltonian, we finally obtained the analytic result

$$\begin{aligned} \boldsymbol{K}^j_{BCS} &= \sum_k \left\{ \epsilon_{bkj+\frac{1}{2}} \boldsymbol{b}^\dagger_{j+\frac{1}{2},k} \boldsymbol{b}_{j+\frac{1}{2},k} + \epsilon_{dkj+\frac{1}{2}} \boldsymbol{d}^\dagger_{j+\frac{1}{2},k} \boldsymbol{d}_{j+\frac{1}{2},k} \right. \\ & \left. + \epsilon_{bkj-\frac{1}{2}} \boldsymbol{b}^\dagger_{j-\frac{1}{2},k} \boldsymbol{b}_{j-\frac{1}{2},k} + \epsilon_{dkj-\frac{1}{2}} \boldsymbol{d}^\dagger_{j-\frac{1}{2},k} \boldsymbol{d}_{j-\frac{1}{2},k} \right\} , \end{aligned} \tag{6}$$

having introduced the notation $\boldsymbol{AB} = \sum_{\lambda cf} \boldsymbol{A}_{\lambda cf} \boldsymbol{B}^{\lambda cf}$. The Hamiltonian is a function in terms of new creation and annihilation operators, where $\boldsymbol{b}^\dagger_{lk\lambda cf}$ describe quarks and $\boldsymbol{d}^\dagger_{lk-\lambda\bar{c}\bar{f}}$ anti-quarks. The numerical part involves the solution of bi-quadratic equations for the unknown variable α^j_{Nk} and β^j_{Nk}. The solution of the bi-quadratic equations is obtained numerically, which is the origin of the notation *semi-analytic*.

The spectrum of (6) is trivial, because only number operators are involved. Some examples for hadronic states are

$$\begin{aligned} & \left[\boldsymbol{b}^\dagger_{j\pm\frac{1}{2},k} \otimes \boldsymbol{d}^\dagger_{j\pm\frac{1}{2},k'} \right]^{J(0,0)_C F}_\mu |BCS\rangle , \\ & \left[\boldsymbol{b}^\dagger_{j\pm\frac{1}{2},k} \otimes \boldsymbol{d}^\dagger_{j\mp\frac{1}{2},k'} \right]^{J(0,0)_C F}_\mu |BCS\rangle , \end{aligned} \tag{7}$$

with the energy $\left(\epsilon_{bkj\pm\frac{1}{2}} + \epsilon_{dkj\pm\frac{1}{2}} \right)$. Because each creation operator represent a complicated sum

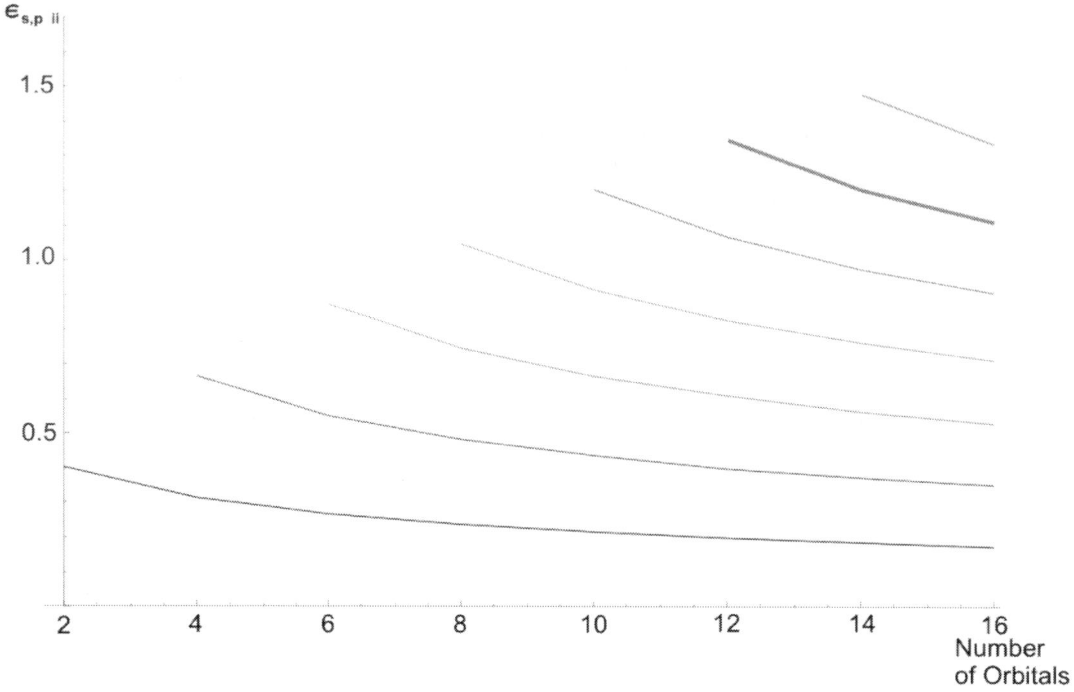

Figure 2. Behavior of the energy levels (GeV) with respect to the number of levels considered.

into the former quark-antiquark operators, they represent a *parton*, where each parton is given by a valence quark (anti-quark) and a sea of quark-antiquarks pairs.

In Fig. 2 the behavior of the lowest one-particle excitations is shown, as a function in the number of orbitals taken into account. The energy of a meson state is then the sum of two one-particle energies. For a large number of orbital levels the lowest meson excitation is about 150 MeV for each parton.

This puts the first meson excitation at 300 MeV. There is still a degeneration related to the coupling to different flavors, i.e., the $F = \frac{1}{2}$, $J^\pi 0^-$ multiplet is degenerate with the $F = \frac{1}{2}$, $J^\pi 1^-$ multiplet. The degeneration will be lifted, once the interaction terms of the Hamiltonian are taken into account.

The diagonalization of the quark part represents an important step towards the complete diagonalization of the QCD Hamiltonian. It represents a *prediagonalization* of the dominant part of the QCD Hamiltonian. Note that it refers to the diagonalization of the *single-particle levels*. Including the interaction terms we will diagonalize again only the single-particle levels. The prediagonalization represents a good chance that the new off-diagonal matrix elements (in the single-particle space) are not too large, as was the case of the kinetic terms of the quarks part), and that a convergence can be reached not inlcuding too many single particle quark and gluon states.

4. Including gluons

In this section we discuss possible methods to include dynamically gluons into the treatment. The discussion will not be complete because it is an on-going investigation. Nevertheless, it will show that standard many-body technique, leaned from nuclear physics [11], may suffice to obtain a complete treatment of QCD at low energy.

Journal of Physics: Conference Series **322** (2011) 012016 doi:10.1088/1742-6596/322/1/012016

4.1. Coherent state method

For the gluons and quarks-antiquarks we use respectively the following coherent states [14, 15]

$$\begin{aligned} \mid \gamma\rangle_g &= \mathcal{N}_g e^{\sum_\nu \gamma_\nu \boldsymbol{\beta}^\dagger_\nu} \mid 0\rangle_g \ , \\ \mid \alpha\rangle_q &= \mathcal{N}_q e^{\sum_\mu \alpha^\mu_a \boldsymbol{b}^\dagger_\mu + \sum_{\bar\mu} \alpha^{\bar\mu}_b \boldsymbol{d}^\dagger_{\bar\mu}} \mid 0\rangle_q \ , \end{aligned} \tag{8}$$

where $\boldsymbol{b}^\dagger_\mu$, $\boldsymbol{d}^\dagger_{\bar\mu}$ and $\boldsymbol{\beta}^\dagger_\nu$ are short-hand notations for the quark-antiquark and gluons respectively, $\bar\mu$ is the adjoint to μ and $\mathcal{N}_g$, $\mathcal{N}_q$ are normalization factors for the gluon and quark part respectively. The variables α^k_μ (k=a, b) are Grassmann variables [15] (in this reference one also can find a detailed description on the mathematical properties).

In (8) care has to be taken in the quark sector. Denoting by $\mathcal{N}$ the total degrees of freedom of the quark sector, the exponential expansion stops at the power of $\mathcal{N}$, i.e., the exponential in (8) in reality are finite sum. Further below we will shortly discuss the error involved, using still the exponential notation.

The expectation of the creation and annihilation operators with respect to these coherent states are

$$\begin{aligned} {}_g\langle \boldsymbol{\beta}^\dagger_\nu\rangle_g &= \gamma_\nu \ , \quad {}_g\langle \boldsymbol{\beta}^\nu\rangle_g = \gamma^\nu \ , \\ {}_q\langle \boldsymbol{b}^\dagger_\mu\rangle_q &= \alpha^a_\mu \ , \quad {}_q\langle \boldsymbol{b}^\mu\rangle_q = \alpha^\mu_a \ , \\ {}_q\langle \boldsymbol{d}^\dagger_{\bar\mu}\rangle_q &= \alpha^b_{\bar\mu} \ , \quad {}_q\langle \boldsymbol{d}^{\bar\mu}\rangle_q = \alpha^{\bar\mu}_b \ . \end{aligned} \tag{9}$$

The expressions in (9) are only approximate. The coherent state for quarks is in reality not an exponential but, with $\boldsymbol{X} = \sum_\mu \alpha^\mu_a \boldsymbol{b}^\dagger_\mu + \sum_{\bar\mu} \alpha^{\bar\mu}_b \boldsymbol{d}^\dagger_{\bar\mu}$,

$$e^{\boldsymbol{X}} \mid 0\rangle = \sum_{n=0}^{\mathcal{N}} \frac{\boldsymbol{X}^n}{n!} \mid 0\rangle \ . \tag{10}$$

Applying a fermion annihilation to it results in a sum from $n = 0$ to $n = (\mathcal{N} - 1)$. The error involved originates from the last term, with $n = \mathcal{N}$, which has to added when we write the final expression in terms of an exponential. The error is proportional to $1/\mathcal{N}!$. Assuming that the variables of the coherent state are of the order of one and that $\mathcal{N}$ is at least 48 (two spin degrees of freedom, two for flavor, three for color and four orbital states, two at positive and two at negative energy), the error involved is negligible.

The next step is to expand the creation and annihilation operators around these expectation values [15, 16], i.e., around a new vacuum,

$$\begin{aligned} \boldsymbol{\beta}^\dagger_\nu &= \gamma_\nu + \delta\boldsymbol{\beta}^\dagger_\nu \ , \quad \boldsymbol{\beta}^\nu = \gamma^\nu + \delta\boldsymbol{\beta}^\nu \ , \\ \boldsymbol{b}^\dagger_\mu &= \alpha^a_\mu + \delta\boldsymbol{b}^\dagger_\mu \ , \quad \boldsymbol{b}^\mu = \alpha^\mu_a + \delta\boldsymbol{b}^\mu \ , \\ \boldsymbol{d}^\dagger_{\bar\mu} &= \alpha^b_{\bar\mu} + \delta\boldsymbol{d}^\dagger_{\bar\mu} \ , \quad \boldsymbol{d}^{\bar\mu} = \alpha^{\bar\mu}_b + \delta\boldsymbol{d}^{\bar\mu} \ . \end{aligned} \tag{11}$$

This mapping conserves the commutation relations of the gluon operators and the anti-commutation relations of the quark operators.

The final step is to apply this mapping to the Hamiltonian and expand this operator up to second order in the creation and annihilation operators. If we denote in general by $\boldsymbol{c}^\dagger_i$ and $\boldsymbol{c}^j$ the creation and annihilation operator for both sectors (*fermion and boson*), fermions and bosons, and exclude third order interaction, the general structure of the Hamiltonian will be

$$\boldsymbol{H} \approx V(\alpha,\gamma) + \sum_{ij} \left(A_{ij}\boldsymbol{c}^\dagger_i \boldsymbol{c}_j + B_{ij}\boldsymbol{c}^\dagger_i \boldsymbol{c}^\dagger_j + B_{ij}\boldsymbol{c}_i \boldsymbol{c}_j \right) \ . \tag{12}$$

Journal of Physics: Conference Series **322** (2011) 012016 doi:10.1088/1742-6596/322/1/012016

This type of Hamiltonian can be solved using BCS and RPA [11] methods. However, in general, also third order interaction terms appear and will need some special attention.

This is as far as we got. In the future we will apply the steps proposed and investigate the possibility to solve the Hamiltonian (12).

4.2. Mixed treatment for quark and gluons

Instead of applying the coherent state method [14] to both sectors, quarks and gluons, one can use it for the gluon sector only. The advantage is that no Grassmann variables have to be dealt with. Also, the gluon sector is the most complicated part, involving large powers in the gluon fields, and a quadratic expansion in $\delta\boldsymbol{\beta}^\dagger_\nu$ and $\delta\boldsymbol{\beta}^\nu$ will simplify the structure of the interacting part. Solving the Hamiltonian as a function in the coherent state parameters γ_ν, will then determine the optimal values for γ_ν minimizing the ground state.

The dynamical part of the gluons are then introduced by expanding around the γ_ν values (see (11), introducing the $\delta\boldsymbol{\beta}^\dagger_\nu$ and $\delta\boldsymbol{\beta}^\nu$ boson operators.

In what follows, we discuss shortly the structure of the pure gluon part (without any mixing to the quark sector), which would be the result of the before mentioned expansion. Expanding the pure gluon sector up to second order terms in $\delta\boldsymbol{\beta}^\dagger_\nu$ and $\delta\boldsymbol{\beta}_\nu$, the structure of the gluon Hamiltonian is as follows

$$\begin{aligned}\boldsymbol{H}_g(\delta\boldsymbol{\beta}^\dagger,\delta\boldsymbol{\beta}) &= \sum_v E_v\delta\boldsymbol{\beta}^\dagger_v\delta\boldsymbol{\beta}^v + \sum_{vw} V_{vw}\delta\boldsymbol{\beta}^\dagger_v\delta\boldsymbol{\beta}^w \\ &+\frac{1}{2}\sum_{vw} W_{vw}\left(\delta\boldsymbol{\beta}^\dagger_v\delta\boldsymbol{\beta}^{\dagger w} + \delta\boldsymbol{\beta}_v\delta\boldsymbol{\beta}^w\right) ,\end{aligned} \tag{13}$$

with v and w is a short hand notation for all possible indices. One possibility is to apply BCS or RPA [11]. For RPA one maps this Hamiltonian to a new one which depends on new boson creation and annihilation operators $\boldsymbol{\Gamma}^\dagger_n$, Γ^n, i.e.,

$$\boldsymbol{H}(\boldsymbol{\Gamma}^\dagger,\boldsymbol{\Gamma}) = \sum_n \Omega_n\left(\boldsymbol{\Gamma}^\dagger_n\boldsymbol{\Gamma}^n + const\right) , \tag{14}$$

with

$$\boldsymbol{\Gamma}^\dagger_n = \sum_v \left[\mathcal{X}_{v,n}\boldsymbol{\beta}^\dagger_v - \mathcal{Y}_{v,n}\boldsymbol{\beta}^v\right] , \tag{15}$$

with $\mathcal{X}_{v,n}$ and $\mathcal{Y}_{v,n}$ being some functions to be determined [11]. This proceedure does nothing else than contructing a new vacuum, due to the effect of the residual interaction, around which a harmonic expansion is applied.

Involving in addition the quark fields require more involved considerations. One possibility is to diagonalize the pure quark sector, as given above, and separately the pure gluon sector, as indicated in this section, and afterwards to include the quark-gluon mixing terms, already expanded up to second order on the gluon fields.

5. Conclusions

Methods to treat QCD at low energies were reviewed. Taking into account only static gluons and restricting to a finite volume, semi-analytic results were obtained, involving an arbitrary number of orbital quark states.

In the next step the gluons will be included dynamically. We reviewed a couple of procedures, with great prospects. They show that probably standard many-body techniques, applied in the right order, will be sufficient to determine the structure of the hadron spectrum at low energy.

Note, that the QCD Hamiltonian, used here, depends only on two parameters, namely the coupling constant g and the size of the volume. The latter can be removed, considering the limit of large volumes.

This gives us hope to possibly derive the structure of phenomenological models in future. This is still an ongoing investigation and the future will show if we are on the right track.

Acknowledgments

We acknowledge financial help from DGAPA and from the National Research Council of Mexico (CONACyT), the CONICET (Argentina), the US. Department of Energy and the help from Mexican Federal Program for the improvement of professors SEP-PROMEP.

References

[1] Lee T D 1981 *Particle Physics and Introduction to Field Theory* (Singapore: World Scientific)
[2] Greiner W, Schramm S and Stein E 2002 *Quantum Chromodynamcis* (Heidelberg: Springer-Verlag)
[3] DeGrand T and DeTar C 2006 *Lattice Methods for Quantum Chromodynamics* (Singapore: World Scientific) PhD-Thesis,
[4] DeGrand T, Jaffe R L, Johnson K and Kiskis J 1975 *Phys. Rev.* D **12** 2060
[5] Bijker R, Iachello F and Leviatan A 1994 *Ann. Phys. (N.Y.)* **236** 69
[6] Lerma S, Jesgarz S, Hess P O, Civitarese O and Reboiro M 2003 *Phys. Rev.* C **67** 055209
Nuñez M V, Lerma S, Hess P O, Jesgarz S, Civitarese O and Reboiro M 2004 *Phys. Rev.* C **70** 035208
[7] Yépez-Martínez T, Hess P O, Szczepaniak A and Civitarese O 2009 *Rev. Mex. Fís.* **55**(2) 60
[8] Yépez-Martínez T, Hess P O, Szczepaniak A and Civitarese O 2010 *Phys. Rev.* C **81** 045204
[9] Szczepaniak A and Swanson E S 2001 *Phys. Rev.* D **65** 025012
[10] Krupinski P 2006 *Gluon Degrees of Freedom in QCD Coulomb Gauge Ha,iltonian* (Indiana University: PhD-Thesis)
Szczepaniak A and Krupinski P 2002 *Phys. Rev.* D **66** 096006
Szczepaniak A and Krupinski P 2006 *Phys. Rev.* D **73** 034022
Szczepaniak A and Krupinski P 2006 *Phys. Rev.* D **73** 116002
[11] Ring P and Schuck P 1980 *The Nuclear Many-Body Problem*, (Heidelberg: Springer)
[12] Escher J and Draayer J P 1998 *J. Math. Phys.* **39** 5123
[13] Draayer J P and Akiyama Y 1973 *J. Math. Phys.* **14** 1904
[14] Hecht K T 1987 *Vector Coherent State Method and its Application to Problems of Higher Symmetries* Lecture Notes in Physics (Heidelberg: Springer-Verlag)
[15] Auerbach A 1994 *Interacting Electrons and Quantum Magnetism* (Heidelberg: Spriner-Verlag)
[16] Klein A and Marshalek E R 1991 *Rev. Mod. Phys.* **63** 375

Journal of Physics: Conference Series **322** (2011) 012017 doi:10.1088/1742-6596/322/1/012017

Nuclear masses, deformations and shell effects

Jorge G Hirsch
Instituto de Ciencias Nucleares, Universidad Nacional Autónoma de México, Apartado Postal 70-543, 04510 México, D.F., Mexico

César A Barbero and Alejandro E Mariano
Departamento de Física, Facultad de Ciencias Exactas, Universidad Nacional de La Plata, CC 67, 1900 La Plata, Argentina

E-mail: hirsch@nucleares.unam.mx

Abstract. We show that the Liquid Drop Model is best suited to describe the masses of prolate deformed nuclei than of spherical nuclei. To this end three Liquid Drop Mass formulas are employed to describe nuclear masses of eight sets of nuclei with similar quadrupole deformations. It is shown that they are able to fit the measured masses of prolate deformed nuclei with an RMS smaller than 750 keV, while for the spherical nuclei the RMS is, in the three cases, larger than 2000 keV. The RMS of the best fit of the masses of semi-magic nuclei is also larger than 2000 keV. The parameters of the three models are studied, showing that the surface symmetry term is the one which varies the most from one group of nuclei to another. In one model, isospin dependent terms are also found to exhibit strong changes. The inclusion of shell effects allows for better fits, which continue to be better in the prolate deformed nuclei region.

1. Introduction

The description of nuclear masses in terms of the Liquid Drop Model paved the way to the basic understanding of nuclear properties, like the saturation of the nuclear force, the existence of pairing and shell effects, and the description of fission and fusion processes [1]. The Q-values of different nuclear reactions, obtained from mass differences, must be accurately known to allow the description of the astrophysical origin of the elements [2]. Accurate theoretical predictions of nuclear masses remain a challenge [3], sharing the difficulties with other quantum many-body calculations, and complicated by the absence of a full theory of the nuclear interaction.

Decades of work have produced microscopic and macroscopic mass formulas [4]. At present, the most successful approaches seem to be the Finite Range Droplet Model (FRDM) [5], the Skyrme and Gogny Hartee Fock Bogolyubov (HFB) [6, 7], and the Duflo-Zuker (DZ) mass formula [8, 9, 10]. They allow for the calculation of masses, charge radii, deformations, and in some cases also fission barriers. They all contain a macroscopic sector which resembles the Liquid Drop Mass (LDM) formula, and include deformation effects. HFB calculations are now able to fit known nuclear masses with deviations competitive with the '95 FRDM calculations, which are also being improved, while the most precise and robust nuclear mass predictions are given by the DZ model [4, 11].

The Liquid Drop Mass (LDM) formula captures the macroscopic features of the mass dependence on the number of neutrons N, of protons Z, and on its mass numbers $A = N + Z$.

It includes volume and surface terms, the Coulomb interaction between protons, Wigner and symmetry terms, linear and quadratic in the neutron excess $N - Z$, and a pairing term. It is generally assumed that the liquid-drop energy of a spherical nucleus is described by a Bethe-Weizsäcker mass formula [12], being a common practice to describe nuclear masses and radii of spherical closed-shell nuclei in terms of a mean field and add deformation and other shell effects as corrections [13].

It is the purpose of this contribution to show that, when nuclei with measured masses are grouped according with their quadrupole deformations, three Liquid Drop Mass formulas consistently allow for a fit with an RMS smaller than 750 keV for the set of most prolate deformed nuclei, while they are unable to fit the masses of spherical nuclei with an RMS smaller than 2000 keV. Semi-magic nuclei are fit with a similarly large RMS. The parameters of the three models are studied, showing that the surface symmetry term is the one which varies the most from one group of nuclei to another. In one model, isospin dependent terms are also found to exhibit strong changes.

Shell effects refer to the differences between the experimental binding energies [14] and the LDM predictions. In this work they are included employing linear, quadratic [15, 16], 3- and 4-body terms [13], functions of the number of valence nucleons, following the ideas of the Duflo-Zuker model [8, 9, 10].The inclusion of shell effects allows for better fits, while the smallest rms are found again in the prolate deformed nuclei region.

2. The three Liquid Drop Mass formulas

Three Liquid Drop Mass formulas will be employed to analyze their ability to fit nuclear masses. One of them treats consistently volume and surface effects [13]. The second one incorporates explicitly isospin effects [12] while the third one includes quadratic isospin effects, the diffuseness correction to the Coulomb energy, the charge exchange correction term and the curvature energy [17].

In Ref. [13] an improved version of the liquid-drop mass formula with modified symmetry and Coulomb terms is built, following a consistent treatment of nuclear bulk and surface effects. The negative nuclear interaction energy is given by:

$$E_{LDM1} = -a_v A + a_s A^{2/3} + S_v \frac{4T(T+1)}{A(1+yA^{-1/3})} + a_c \frac{Z(Z-1)}{(1-\Lambda)A^{1/3}} - a_p \frac{\Delta}{A^{1/3}}, \tag{1}$$

where the pairing interaction is given by $\Delta = 2$, 1, and 0 for even-even, odd-mass and odd-odd nuclei, respectively. A modification Λ in the Coulomb term is included

$$\Lambda = \frac{N-Z}{6Z(1+y^{-1}A^{1/3})} \tag{2}$$

Its presence in the denominator of the Coulomb term suggests that it can be viewed as a correction to the radius of the nucleus. The symmetry term employs $4T(T+1)$ instead of $(N-Z)^2$ to account for the Wigner energy. The Coulomb interaction is proportional to $Z(Z-1)$ to avoid the Coulomb interaction of a proton with itself.

The second LDM formula we are going to analyze was introduced in Ref. [12] by considering isospin effects. In this case the liquid-drop energy of a spherical nucleus is described by a modified Bethe-Weizsäcker mass formula

$$E_{LDM2} = -a_v A + a_s A^{2/3} + a_{sym} I^2 A + a_c \frac{Z(Z-1)}{A^{1/3}}(1 - Z^{2/3}) - a_{pair} \frac{\delta_{np}}{A^{1/3}}, \tag{3}$$

Journal of Physics: Conference Series **322** (2011) 012017 doi:10.1088/1742-6596/322/1/012017

with isospin asymmetry $I = (N - Z)/A$. The pairing term is taken form [18]

$$\delta_{np} = \begin{cases} 2- \mid I \mid & \text{:N and Z even} \\ \mid I \mid & \text{:N and Z odd} \\ 1- \mid I \mid & \text{:N even, Z odd, and N>Z} \\ 1- \mid I \mid & \text{:N odd, Z even, and N<Z} \\ 1 & \text{:N even, Z odd, and N<Z} \\ 1 & \text{:N odd, Z even, and N>Z.} \end{cases} \tag{4}$$

and the symmetry energy coefficient of finite nuclei is written as,

$$a_{sym} = c_{sym}\left[1 - \frac{\kappa}{A^{1/3}} + \frac{2- \mid I \mid}{2+ \mid I \mid A}\right] \tag{5}$$

based on the conventional surface-symmetry term of liquid-drop model, with a small correction term for description of isospin dependence of a_{sym}. The symmetry energy coefficient a_{sym} increases with decreasing isospin asymmetry. This I correction term approximately describes the Wigner effect for heavy nuclei.

Different mass formulae derived from the liquid drop model and the pairing and shell energies of the Thomas-Fermi model have been studied and compared in Ref [17]. We selected for this study to include the diffuseness correction to the Coulomb energy, the charge exchange correction term and the curvature energy. In Ref. [17] it is reported that the Coulomb diffuseness correction Z^2/A term and the charge exchange correction $Z^{4/3}/A^{1/3}$ term play the main role to improve the accuracy of the mass formula. The Wigner term and the curvature energy can also be used separately but their coefficients are very unstable.

Their LDM formula is

$$\begin{aligned} E_{LDM3}(N,Z) = \; & -a_v(1-k_vI^2)A + a_s(1-k_sI^2)A^{2/3} + a_k(1-k_kI^2)A^{1/3} + \frac{3}{5}\frac{e^2Z^2}{r_0A^{1/3}} \\ & -f_pZ^2/A - a_{c,exc}Z^{4/3}/A^{1/3} + E_{pair} \end{aligned} \tag{6}$$

The volume energy corresponding to the saturated exchange force and infinite nuclear matter is given by the first term. I^2A is the asymmetry energy of the Bethe-Weizsäcker mass formula. The second term is the surface energy. Its origin is the deficit of binding energy of the nucleons at the nuclear surface and corresponds to semi-infinite nuclear matter. The following term is the curvature energy. It results from non-uniform properties which correct the surface energy and depends on the mean local curvature. The decrease of binding energy due to the Coulomb repulsion is given by the fourth term, which has and adjustable charge radii $r_0A^{1/3}$. The Z^2/A term is the diffuseness correction to the sharp radius Coulomb energy, also called also the proton form-factor. The $Z^{4/3}/A^{1/3}$ term is the charge exchange correction term. The pairing energies of the Thomas-Fermi model [19] were employed.

3. The fits

The coefficients of the three LDM were selected to minimize the root mean square deviation (RMS) when the predicted binding energies $BE_{\text{th}}(N,Z)$ are compared with the experimental ones $BE_{\text{exp}}(N,Z)$, reported in AME03 [14], modified so as to include more realistically the electron binding energies as explained in Appendix A of Lunney, Pearson and Thibault [4].

$$\text{RMS} = \left\{\frac{\sum\left[BE_{\text{exp}}(N,Z) - BE_{\text{th}}(N,Z)\right]^2}{N_{nucl}}\right\}^{1/2}. \tag{7}$$

N_{nucl} is the number of nuclei in each group, listed in the fourth row of Table 1. The minimization procedure uses the routine Minuit [20].

The fits were performed employing the masses of nine groups of nuclei:

Journal of Physics: Conference Series **322** (2011) 012017 doi:10.1088/1742-6596/322/1/012017

Table 1. The nine groups on nuclei employed in the present study, their range of quadrupole deformation, and their number of nuclei.

group	all	1	2	3	4	5	6	7	semi-magic
ϵ_2 min	-0.65	-0.65	-0.11	0.00	0.04	0.12	0.18	0.23	
ϵ_2 max	0.65	-0.11	0.00	0.04	0.12	0.18	0.23	0.65	
N_{nucl}	2149	258	252	332	272	307	364	364	185

- all nuclei whose measured masses are reported in AME03 [14], which have N, $Z \geq 8$,
- seven groups of nuclei whose quadrupole deformations ϵ_2, taken form the FRDM [21], lie in the ranges listed in the second and third row of Table 1,
- the group of all semi-magic nuclei, having Z= 8, 20, 28, 50, 82 or N= 8, 20, 28, 50, 82 or 126.

Notice that group 1 contains most of the oblate nuclei, that the more spherical nuclei belong to group 3, and that the more prolate deformed nuclei are included in groups 6 and 7.

For each LDM equation, nine fits were performed, one for each group of nuclei. In this way, nine sets of parameters were obtained, which minimize the RMS of each group of nuclei. Employing these nine sets of parameters, the RMS were estimated for all groups, whose RMS are shown in the next subsections.

3.1. Analysis of the LDM1 formula

Here we show the results obtained using Eq. (1) for the nine regions. In our calculations we select nine sets of fixed values for the parameters a_v, a_s, a_c, a_p, S_v and y. The results are exhibited in Tables 2 and 3.

Table 2. RMS (in keV) for the nine groups (columns), for the nine sets of parameters (rows), employing Eq. (1)

	all	1	2	3	4	5	6	7	semi-magic
set_{all}	**2387**	1842	3814	3080	2037	1495	1815	2060	3888
set_1	3300	**1313**	3113	3420	3277	3173	5194	1476	4016
set_2	4053	2917	**1676**	2433	3088	4273	6243	4701	2570
set_3	3404	3313	2298	**2063**	2556	3408	4016	4721	2600
set_4	2567	2431	3867	3096	**1746**	1741	1975	2626	3901
set_5	2630	2097	4848	3696	2072	**1053**	1360	1735	4521
set_6	2828	1942	5205	4262	2578	1469	**870**	1296	5078
set_7	3169	1655	4077	4220	3543	2753	3643	**746**	5016
set_{semi}	4514	3816	1930	2665	3316	4777	6729	5306	**2113**

3.2. Analysis of the LDM2 formula

Here we show the results obtained using Eq. (3) for the nine regions. In our calculations we select nine sets of fixed values for the parameters a_v, a_s, a_c, a_{pair}, c_{sym} and κ. The results are exhibited in Tables 4 and 5.

Table 3. Sets of parameters which minimize the RMS for the nine groups of nuclei, employing Eq. (1). In the last three rows the average value of each parameter, their dispersion and percentage variation $\%\mathrm{var} = 100\frac{\sigma}{|\mathrm{average}|}$ are shown.

	a_v	a_s	a_c	a_p	S_v	y
set_{all}	15.822	18.491	0.703	6.048	30.701	2.616
set_1	15.515	17.577	0.686	6.092	25.997	1.583
set_2	15.883	18.622	0.712	5.733	28.711	2.284
set_3	15.978	18.884	0.716	5.895	32.091	3.059
set_4	15.879	18.564	0.708	4.103	31.459	2.666
set_5	15.832	18.435	0.707	5.266	30.575	2.482
set_6	15.609	17.835	0.689	5.334	29.762	2.388
set_7	15.553	17.782	0.684	5.126	27.402	1.865
set_{semi}	15.775	18.012	0.712	5.500	26.407	1.489
average	15.761	18.245	0.702	5.455	29.234	2.270
σ	0.153	0.426	0.012	0.579	2.093	0.495
% var	1.0	2.3	1.7	10.6	7.1	21.8

Table 4. RMS (in keV) for the nine groups (columns), for the nine sets of parameters (rows), employing Eq. (3)

	all	1	2	3	4	5	6	7	semi-magic
set_{all}	**2374**	1788	3733	3081	2044	1502	1855	2058	3916
set_1	3208	**1254**	3114	3400	3243	3077	4947	1435	4060
set_2	4092	2965	**1675**	2380	3027	4318	6393	4697	2380
set_3	3475	3352	2173	**2069**	2596	3518	4183	4816	2551
set_4	2561	2408	3761	3052	**1762**	1771	2036	2688	3897
set_5	2616	2073	4823	3702	2083	**1021**	1311	1704	4548
set_6	2795	1865	5130	4234	2619	1435	**838**	1242	5090
set_7	3074	1589	4216	4238	3541	2590	3138	**656**	5093
set_{semi}	4677	3915	1959	2722	3427	4953	7036	5466	**2056**

3.3. Analysis of the LDM3 formula

Here we show the results obtained using Eq. (6) for the nine regions. In our calculations we select nine sets of fixed values for the parameters a_v, a_s, r_0, a_{pair}, k_v, k_s, f_p, $a_{c,exc}$, a_k andk_k. The results are exhibited in Tables 6 and 7.

3.4. LDM and deformation

Tables 2, 4, 6 display the RMS obtained with the three LDM formulas. Each row refers to one fixed set of parameters, each column to one group of nuclei.

For the LDM1, Table 2, the global RMS is 2.39 MeV for the 2149 nuclei. In some groups, containing around two or three hundred nuclei, the RMS obtained with this set of parameters is smaller than the global one. The largest RMS are found in groups 2 and 3, those containing spherical nuclei. Along each column, corresponding to one group of nuclei, the smallest RMS

Table 5. Sets of parameters which minimize the RMS for the nine groups of nuclei, employing Eq. (3). In the last three rows the average value of each parameter, their dispersion and percentage variation %var $= 100\frac{\sigma}{|\text{average}|}$ are shown.

	a_v	a_s	a_c	a_{pair}	c_{sym}	κ
set_{all}	15.711	18.920	0.720	6.989	30.045	1.587
set_1	15.432	18.119	0.700	6.793	26.488	1.260
set_2	15.695	18.802	0.722	7.443	27.979	1.441
set_3	15.903	19.395	0.737	6.717	30.582	1.642
set_4	15.729	18.877	0.721	4.685	30.858	1.635
set_5	15.685	18.756	0.721	6.210	29.969	1.548
set_6	15.493	18.254	0.703	6.276	29.490	1.553
set_7	15.500	18.398	0.701	6.193	27.715	1.361
set_{semi}	15.627	18.329	0.724	5.327	26.043	1.099
average	15.642	18.650	0.717	6.293	28.797	1.458
σ	0.138	0.383	0.012	0.801	1.687	0.175
% var	0.9	2.0	1.7	12.7	5.8	12.0

Table 6. RMS (in keV) for the nine groups (columns), for the nine sets of parameters (rows), employing Eq. (6)

	all	1	2	3	4	5	6	7	semi-magic
set_{all}	**2422**	1884	3773	3186	2031	1519	1861	2125	3937
set_1	3576	**1183**	3067	3785	3461	3456	5854	1283	4424
set_2	4211	3032	**1597**	2648	2928	4409	6596	4871	2333
set_3	3563	3398	2197	**2151**	2618	3628	4455	4808	2499
set_4	3049	3035	3849	4090	**1517**	1764	2560	3456	3952
set_5	2776	2406	4838	4004	1927	**986**	1587	2110	4464
set_6	2895	2083	5073	4459	2701	1471	**819**	1579	5103
set_7	3200	1484	4128	4317	3506	2701	3790	**629**	5135
set_{semi}	4707	3776	1781	2720	3337	4962	7321	5392	**1967**

always corresponds the set of parameters obtained fitting in this group, as expected. These RMS values are displayed in bold numbers. It is remarkable that the smallest RMS values are found in the two groups having the more oblate deformed nuclei, in groups 6 and 7, with 0.87 and 0.75 MeV, respectively. On the other hand, the best fit of the spherical nuclei in group 3 has an RMS larger than 2.0 MeV, and those nuclei with very small quadrupole deformation, belonging to groups 2 and 4, have RMS larger than 1.6 MeV. Consistently, the semi-magic group of nuclei, which have small quadrupole deformations, have an RMS of 2.1 MeV.

By analyzing the RMS for each column, it is possible to notice that the smaller RMS are always found around the smallest one. It supports the idea that the division in the seven groups with different deformations makes sense, because the parameters obtained fitting nuclei with a close deformation produce also a small RMS. Notice, for example, that among the RMS of groups 1 and 7, the more oblate and prolate deformed, respectively, the largest RMS are found with the sets of parameters 2 and 3, i.e. those fitted for spherical nuclei. On the opposite

Table 7. Sets of parameters which minimize the RMS for the nine groups of nuclei, employing Eq. (6). In the last three rows the average value of each parameter, their dispersion and percentage variation %var = $100\frac{\sigma}{|\text{average}|}$ are shown.

	a_v	a_s	r_0	a_{pair}	k_v	k_s	f_p	$a_{c,exc}$	a_k	k_k
set$_{all}$	15.647	20.610	1.198	−1.002	2.079	4.451	1.923	0.296	−4.738	30.717
set$_1$	15.834	21.074	1.203	−0.905	1.305	−0.812	1.265	0.304	−5.910	−12.666
set$_2$	15.557	20.860	1.213	−1.037	1.954	4.339	2.583	0.418	−4.719	32.562
set$_3$	15.673	19.718	1.234	−1.005	1.734	2.072	2.071	0.394	−1.248	21.077
set$_4$	15.671	22.448	1.227	−0.726	2.720	9.059	2.936	0.661	−7.069	58.552
set$_5$	15.577	21.060	1.196	−0.873	2.398	6.963	2.259	0.334	−6.359	44.965
set$_6$	15.611	20.997	1.158	−0.963	2.403	6.781	1.457	0.084	−8.514	33.205
set$_7$	15.806	20.834	1.204	−0.962	1.393	−0.390	1.393	0.289	−4.988	−12.233
set$_{semi}$	15.363	18.242	1.221	−0.904	1.418	0.666	2.577	0.211	0.538	−0.408
average	15.638	20.649	1.206	−0.931	1.934	3.681	2.052	0.332	−4.779	21.752
σ	0.132	1.077	0.021	0.089	0.479	3.313	0.558	0.149	2.662	23.694
% var	0.8	5.2	1.7	9.5	24.8	90.0	27.2	44.9	55.7	108.7

side, along columns 3 and 4, containing the more spherical nuclei, the largest RMS are found employing the sets 1 and 7, and in some cases 2 and 6.

It is tempting to conclude that the LDM is best suited for the description of prolate quadrupole deformed nuclei. In order to find support for this conclusion, it is worth to analyze the results presented in Tables 4 and 6. The global RMS for LDM2 is 2.37 MeV, and for LDM3 is 2.42 MeV. The three fits are pretty close to each other. The smallest RMS are always found for the 364 nuclei belonging to group 7, the more prolate deformed, with 0.65 MeV and 0.63 MeV for LDM2 and LDM3, respectively. The masses of the 332 spherical nuclei included in group 3, and the 185 semi-magic nuclei can hardly be fitted with and RMS smaller that 2.0 MeV. Also the correlations between groups are similar for these two other models.

This is the most relevant result reported in this contribution: **the Liquid Drop Model is best suited to describe the masses of prolate deformed nuclei than of spherical nuclei.**

3.5. Comparison of the three LDM formulas

While being close to each other, the three LDM formulas employed in this work have differences which are worth to be studied in detail. A statistical analysis of the parameters of each model probed to be useful in previous studies [11]. Employing the nine values of each parameter, their average, fluctuations and percentage fluctuations are presented in the last three rows of Tables 3, 5 and 7.

The LDM1, Eq. (1) and LDM2, Eq. (3) are very similar. Their volume, surface and Coulomb parameters, listed in Tables 3 and 5, have nearly equal numerical values, and are all of them very stable, with a dispersion along the nine sets smaller than 2%. The pairing parameter fluctuate between 10% and 13%, and the coefficient of the symmetry term around 6-7%. The surface symmetry term y in LDM1 is the most unstable, with fluctuations of 22%, while its counterpart in LDM2, κ, has fluctuations of 12%. In this subtle sense, LDM2 could be considered more stable than LDM1. It could be useful to perform a deeper comparison of the surface symmetry terms, which in one model was obtained asking for consistency between the volume and surface contributions, while in the other was designed to incorporate the isospin dependence explicitly.

The LDM3, Eq. (6), has also stable volume and Coulomb (charge radius) terms, but the

Journal of Physics: Conference Series **322** (2011) 012017 doi:10.1088/1742-6596/322/1/012017

surface term has fluctuations of the order of 5%, as seen in Table 7. The pairing parameter, as in the other two models, fluctuates around 10%. On the other hand, the model has ten parameters, and the remaining six have enormous fluctuations in both magnitude and sign. These instabilities of the model parameters could be interpreted as a weakness which should be addressed. They could be related with redundancies of various terms, allowing for different combinations with similar RMS.

4. Shell effects

The main obstacle for an accurate description of spherical nuclei employing a Liquid Drop Mass formula are the shell effects around closed shells.

In the literature many different ways of implementing shell corrections to the LDM can be found; in general, these methods are rather laborious. A simple method was proposed in Refs. [15, 16] based on counting the number of valence nucleons. This shell correction is linear and quadratic in the total number of valence nucleons n and z,

$$E_{LDMM} = E_{LDM1} + b_1(n+z) + b_2(n+z)^2 \tag{8}$$

where n and z are the numbers of valence neutrons and protons (particle- or hole-like) and b_i are parameters. Inclusion of these two terms in the LDM mass formula (1) reduces the rms deviation from 2.39 to 1.05MeV.

Following Ref. [13], we employ also an upgraded version of the terms (8), which is suggested by the microscopic mass formula of Duflo and Zuker [10, 22]:

$$E_{LDMM'} = E_{LDMM} + a_1 S_2 + a_2 (S_2)^2 + a_3 S_3 + a_{np} S_{np} \tag{9}$$

where

$$S_2 = \frac{n\bar{n}}{D_n} + \frac{z\bar{z}}{D_z}, \quad S_3 = \frac{n\bar{n}(n-\bar{n})}{D_n} + \frac{z\bar{z}(z-\bar{z})}{D_z}, \quad S_{np} = \frac{n\bar{n}}{D_n}\frac{z\bar{z}}{D_z}, \tag{10}$$

with $\bar{n} = D_n - n$ and $\bar{z} = D_z - z$, where $D_n(D_z)$ is the degeneracy of the neutron (proton) valence shell.

They include 2-, 3- and 4-body terms. The quadratic term is associated to configuration mixing and the cubic one to a genuine three body force [18].

In Table 8 the RMS of the best fits for all the nuclei, for the seven sets of nuclei grouped according to their deformations, and for the semi-magic nuclei, are presented for the Liquid Drop Model, Eq. (1), for the Modified Liquid Drop Model, Eq. (8) and for the model including 3- and 4-body terms, Eq. (9). The first row corresponds to the RMS listed in Table 2 in bold face numbers. It is clear that the inclusion of microscopic terms improves the fits. The global RMS, for all nuclei, diminished form its LDM value of 2.39 MeV to 1.07 MeV and 0.89 MeV. The most impressive reductions in the RMS are found in the spherical nuclei grouped in region 3, which drops from 2.06 MeV to 1.01 MeV and 0.90 MeV, and for the semi-magic nuclei, whose RMS diminishes from 2.11 MeV to 1.04 MeV and 0.82 MeV.

On the other hand, the nuclei in region 7, the most prolate deformed, are the best fitted in the LDMM, while the spherical nuclei in region 3 and the semi-magic nuclei have the largest RMS. The inclusion of 2-, 3- and 4-body terms in LDMM' seems to succeed in introducing deformation effects. Regions 1 and 4 to 7 have all RMS between 562 and 623 keV. Spherical and semi-magic nuclei remain to be those with the largest RMS.

Table 8. RMS (in keV) of the best fit for each of the nine groups (columns), employing Eqs. (1), (8) and (9), respectively.

	all	1	2	3	4	5	6	7	semi-magic
LDM1	2387	1313	1676	2063	1746	1053	870	746	2113
LDMM	1075	797	962	1007	828	711	792	616	1038
LDMM'	888	623	741	902	634	562	620	575	817

5. Conclusions

Along this contribution we have shown that the Liquid Drop Model is best suited to describe the masses of prolate deformed nuclei than of spherical nuclei. The analysis was performed employing three different Liquid Drop Mass formulas. With them, the nuclear masses of nuclei grouped in eight sets with similar quadrupole deformations were fitted. For the three LDM models it was found that the masses of prolate deformed nuclei can be described with remarkable precision for a LDM, with an RMS smaller than 750 keV, while the masses of spherical and semi-magic nuclei are those worst described, with RMS larger than 2000 keV.

The dispersion of the parameters of the three models were studied comparing the fits for the different groups of nuclei. We found that in the three the surface symmetry term is the one which varies the most from one group of nuclei to another. In the model of Ref. [17], isospin dependent terms were found to exhibit strong changes, making this model the least robust of the three under this criterion.

The inclusion of shell effects allows for better fits, which continue to be better in the prolate deformed nuclei region. The Duflo-Zuker model is based in a microscopic description of shell effects, and describes deformation through a change in valence occupations. It remains a challenge to see if the DZ mechanism to incorporate deformation effects can be successfully employed by other models.

Acknowledgments

C.B. and A.M. are fellows of CONICET. This work was supported in part by Conacyt, México, by FONCICYT project 94142, by DGAPA-UNAM, and by the Agencia Nacional de Promoción Científica y Tecnológica, Argentina, through PICT-2007-00861..

References

[1] Bohr A and Mottelson B R 1998 *Nuclear Structure vol I* (Singapore: World Scientific)
[2] Rolfs C E and Rodney W S 1988 *Cauldrons in the Cosmos* (Chicago: University of Chicago Press)
[3] Blaum K 2006 *Phys. Rep.* **425** 1
[4] Lunney D, Pearson J M and Thibault C 2003 *Rev. Mod. Phys.* **75** 1021
[5] Möller P, Nix J R, Myers W D and Swiatecki W J 1995 *At. Data Nucl. Data Tables* **59** 185
[6] Goriely S, Tondeur F and Pearson J M 2001 *At. Data Nucl. Data Tables* **77** 311
Goriely S, Samyn M and Pearson J M 2007 *Phys. Rev. C* **75** 064312
[7] Goriely S, Chamel N and Pearson J M 2009 *Phys. Rev. Lett.* **102** 152503
Goriely S, Hilaire Girod S M and Péru S 2009 *Phys. Rev. Lett.* **102** 242501
[8] Duflo J 1994 *Nucl. Phys. A* **576** 29
[9] Zuker A P 1994 *Nucl. Phys. A* **576** 65
[10] Duflo J and Zuker A P 1995 *Phys. Rev. C* **52** R23
[11] Mendoza-Temis J, Morales I, Barea J, Frank A, Hirsch J G, López-Vieyra J C, Van Isacker P and Velázquez V 2008 *Nucl. Phys. A* **812** 28
[12] Wang N, Liu M and Wu X 2010 *Phys. Rev. C* **81** 044322
[13] Dieperink A E L and Van Isacker P 2009 *Eur. Phys. J. A* **42** 269
[14] Audi G, Wapstra A H and Thibault C 2003 *Nucl. Phys. A* **729** 337

[15] Dieperink A E L and Van Isacker P 2007 *Eur. Phys. J. A* **32** 11
[16] Mendoza-Temis J, Frank A, Hirsch J G, López-Vieyra J C, Morales I, Barea J, Van Isacker P and Velázquez V 2008 *Nucl. Phys. A* **799** 84
[17] Royer G, Guilbaud M and Onillon A 2010 *Nucl. Phys. A* **847** 24
[18] Mendoza-Temis J, Hirsch J G and Zuker A P 2010 *Nucl. Phys. A* **843** 14
[19] Myers W D and Swiatecki W J 1966 *Nucl. Phys. A* **601** 141
[20] James F 1994 *Minuit: Function Minimization and Error Analysis Reference Manual, Version 94.1, CERN* ⟨ http://wwwasdoc.web.cern.ch/wwwasdoc/minuit/minmain.html ⟩
[21] Möller P, Nix J R, Myers W D and Swiatecki W J 1995 *At. Data Nucl. Data Tables* **59** 185
[22] Zuker A P 2008 *Rev. Mex. Fís. S* **54** 129 ⟨ http://rmf.fciencias.unam.mx/pdf/rmf-s/54/3/54_3_129.pdf ⟩

On prolate shape predominance in nuclear deformation

P Stránský[1,2], A Frank[1] and R Bijker[1]

[1] Instituto de Ciencias Nucleares, Universidad Nacional Autónoma de México, Apartado Postal 70-543, 04510 México, D.F., Mexico
[2] Institute of Particle and Nuclear Physics, Faculty of Mathematics and Physics, Charles University, V Holešovičkách 2, 180 00 Prague, Czech Republic

E-mail: pavel.stransky@nucleares.unam.mx

Abstract. We use the volume-conserving deformed liquid drop model to perform a quantitative study of the experimentally observed strong predominance of prolate over oblate shapes of nuclei. Except for very light nuclei, the binding energy difference between both types of quadrupole deformation is given by almost equal contributions from the Coulomb and surface terms. We perform an analysis using up-to-date experimental data for electric quadrupole moments and $B(E2)$ transition probabilities and show that, for a significant number of well-deformed nuclei, the prolate-oblate difference in binding energy reaches values of the order of 500-800 keV, which are considerably higher than typical low-lying excitation energies for these nuclei.

1. Introduction

Why do the majority of deformed nuclei have an elongated, cigar-like prolate shape, while there are only few flattened, oblate nuclei? This insistent question has attracted the attention of nuclear physicists for more than half a century. A definitive explanation for the origin of this preponderance is still a matter of discussion and represents a fundamental challenge in nuclear structure physics [1, 2]. The fact that prolate shapes are predominant is evident from a detailed analysis of intrinsic electric quadrupole moments obtained by experiment [3] as displayed in Figure 1, where the sign of the quadrupole deformation parameter β, extracted from measured electric quadrupole moments, determines the nuclear shape: $\beta > 0$ corresponds to prolate, $\beta = 0$ to spherical, and $\beta < 0$ to oblate deformation. It is obvious from the Figure that most of the nuclei lie in the half-plane $\beta > 0$, while oblate shapes are quite exceptional.

Due to the fundamental nature of this phenomenon, a large number of attempts have been made over the years to uncover its physical origin. The microscopic onset of deformation is accounted for by the Federman-Pittel mechanism, as due to the attractive interaction between protons and neutrons occupying orbitals which are spin-orbit partners and have a large radial overlap [4]. This mechanism, however, does not in itself distinguish between prolate and oblate configurations. Most of the models that propose explanations for the prolate-oblate shape assymetry are based on single-particle Nilsson-type shell models with a deformed harmonic oscillator potential.

Ref. [5] presents a qualitative illustration relying on the fact that in the Nilsson scheme, downward (upward) sloping orbits of a given spin I have low (high) values of the projection K onto the symmetry axis on the prolate side, while the opposite is true on the oblate side. In a major shell with many different I-shells there are more orbits with low K, whose mutual

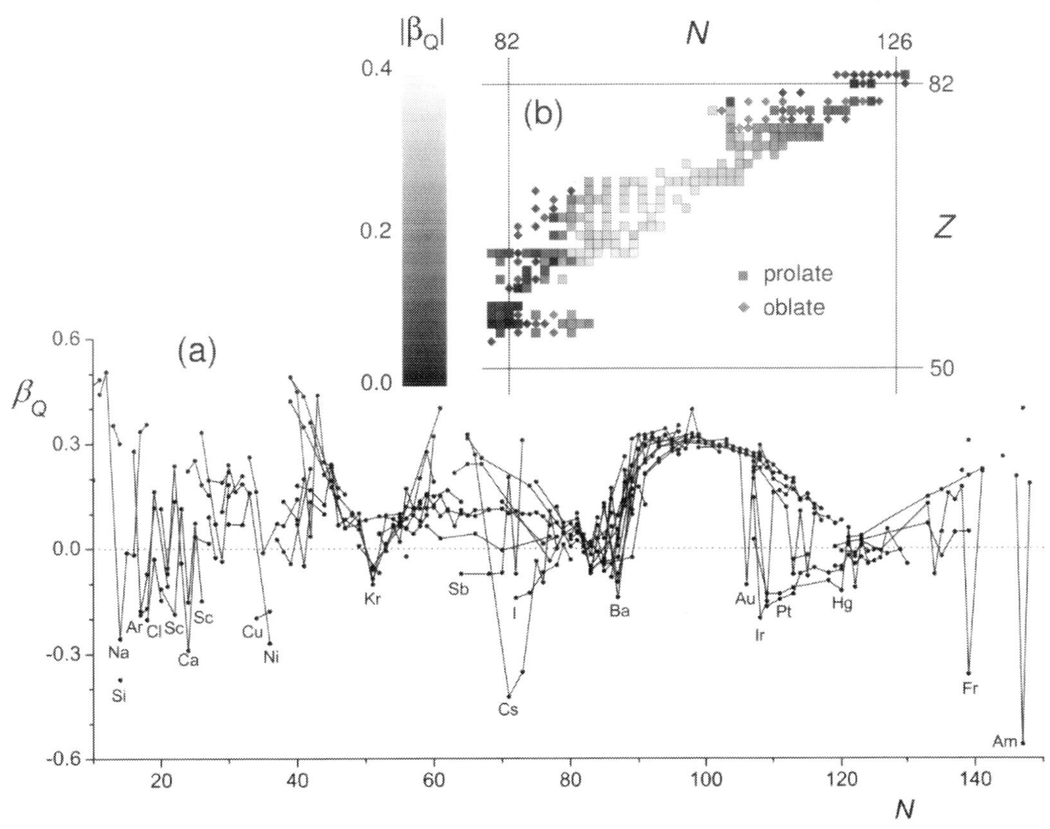

Figure 1. Deformation parameter β_Q calculated from the electric quadrupole moments Q. Most nuclei have prolate shape. (a) Isotopes of each element are connected by lines. Nuclei with significantly high oblate deformation are labeled by their chemical symbols. (b) A detail of the rare-earth region nuclei. The deformation size is coded in shades of gray, and the sign is expressed by squares (prolate shape) and diamonds (oblate shape). It is obvious that a majority of oblate nuclei appears mostly near shell closures.

interaction forces the lowest prolate orbits even deeper in energy. The related gain in binding energy allows to obtain a prolate preference in the beginning of a shell and leads towards oblate shapes later. However, this bulk effect is sensitive to the choice of a potential and strength of spin-spin and spin-orbit couplings [6].

Concerning quantitative studies, the simplest version of the model, without any residual interaction and without the spin-orbit, l^2 and pairing terms, gives an equal number of prolate (until a valence shell is half-filled with nucleons) and oblate shapes (from half-filled to filled shells), although the deformation is slightly larger on the prolate side [1, 7, 8]. Assuming a sharply defined nuclear surface in single-particle models, in order to approach the more realistic Wood-Saxon potential, leads to (i) a shell model with an isotropic quartic correction [7], and (ii) a model of particles bound in an infinite-well ellipsoidal cavity [1]. Both models show (for numbers of particles relevant for nuclear physics) that a prolate shape is energetically favorable, with its energy usually a few MeV lower than a corresponding oblate shape. However, these models predict that oblate states prevail mainly at the beginning of shells and prolate at the end, which is at odds with the experimental observation (see Figure 1). Prolate shape preponderance in the infinite-well ellipsoidal cavity has also been examined from a semiclassical point of view by means of an analysis of classical periodic orbits [9].

Recently, Horoi and Zelevinsky [2] carried out a statistical study of the shell model with a random two-body interaction, with angular momentum, parity and isospin conservation. Their analysis shows that for a subset of realizations of nuclei exhibiting collective features carefully selected from the whole ensemble of all possible realizations, the ratio of prolate shapes can reach 80%, depending drastically, however, on the choice and order of interacting levels; for other subsets oblate forms can predominate.

Explanations in terms of shell effects in Nilson-type models or schematic non-interacting particle models are seemingly in conflict with the robustness of the predominance of prolate over oblate nuclei and the fact that deformation is a collective phenomenon, rather than a single-particle effect. Although compelling, most of them are of a qualitative nature, are model dependent, or they need some fine-tuning to achieve the required result.

In contrast to the single-particle methods, explanations based on collective approaches have also been put forward. Zickendraht [10] demonstrates that the volume element measuring available configuration space is larger for the prolate shape than for the oblate. Moments of inertia for prolate and oblate deformations have been shown to differ [11]. Collective models with single-particle (Strutinsky) corrections have been studied in applications to nuclear masses [12], giving predominance of prolate shapes. In addition, it has been pointed out that the prolate-oblate energy difference is relevant in determining the low-lying structure of even-even nuclei [11, 13].

Incorporating the spin-orbit, l^2 and pairing interactions in the shell model with deformed harmonic oscillator potential, and applying the Nilsson-Strutinsky method (containing a macroscopic collective part) to determine the shape, analyses the problem from both the macroscopic and microscopic points of view. Results of the calculation give the ratio of prolate to all nuclei varying from 40% to 89% [14] depending on the interaction strengths, which are usually separately fixed by experiment for individual shells [12]. This is in agreement with experimental observations, but being an emergent property of the model coupling the microscopic and collective parts, it impedes a transparent discussion of the physical origin of the phenomenon of the prolate shapes preponderance. Similar conclusions can be extracted from other sophisticated numerical models such as Hartree-Fock self-consistent calculations [15]. As Wigner famously quipped: "I am very glad that the computer understands the problem, but I would like to understand it too" [16].

In this paper we aim to demonstrate that a simple form of the collective model of nuclear masses, the deformed Liquid Drop Model (LDM), gives rise to a significant energy difference between prolate and oblate shapes, originating from a sum of surface and Coulomb energy contributions, and thus offers a robust explanation for the observed preponderance of prolate over oblate nuclei. Although the contribution from Coulomb repulsion, which favors an elongated shape, is mentioned in the literature [1, 9], it is usually dismissed as being too small to be a significant effect and relevant only for heavy nuclei. We stress the importance of the surface term which, to the best of our knowledge, has been overlooked in earlier studies, but which gives a crucial contribution to the prolate-oblate energy difference, especially for lighter nuclei. It is known that the Bohr-Mottelson β deformation variable takes values close to 0.3 for well-deformed nuclei, which is sufficient to evaluate the order of magnitude of prolate-oblate collective energy difference. However, for the sake of completeness, we carry out a careful analysis of available experimental data on quadrupole moments [3] and $B(E2)$ transition probabilities [17] to determine the prolate-oblate preponderance in detail. We show that for a significant number of nuclei, the collective part predicts the prolate configuration to be 500-800 keV lower in energy that the oblate one.

2. Deformed Liquid Drop Model

The deformed liquid drop model was first formulated by Von Weizsäcker and Bethe [18]. It treats the atomic nucleus as an incompressible charged fluid of nuclear matter with uniform density and a sharp surface. Its total energy is given by the difference of the mass of N neutrons and Z protons and the binding energy:

$$B_{\mathrm{LQM}} = a_V A - a_S A^{\frac{2}{3}} f_S(\alpha) - a_C \frac{Z(Z-1)}{A^{\frac{1}{3}}} f_C(\alpha) - a_A \frac{(A-2Z)^2}{A} + \delta\ . \tag{1}$$

where $A = N + Z$ is the mass number. The volume term (a_V) expresses the saturation of short-range nuclear interaction. It is corrected by the surface term (a_S), reflecting the absence of interaction between surface nucleons and the exterior of the nucleus. Electrostatic repulsion between protons further diminishes the bond via the Coulomb term (a_C). An imbalance between protons and neutrons gives rise to the asymmetry term (a_A), and spin-coupling to a pairing term (δ). All the parameters $a_{(\bullet)}$ of the model are positive [18, 19, 20].

Since its original publication, this model has been improved by including new degrees of freedom, such as a compressibility and polarizability of the nuclear fluid, a diffuse surface layer, various higher multipole deformations and additional effects related to the deformation (higher order curvature terms), as well as terms augmenting the original macroscopic model with features that come from the microscopic theory, such as shell corrections and an improved pairing contribution. At present, this so-called droplet model and its further extensions are commonly used as a starting point for fitting and predicting nuclear masses and studying nuclear fission [12, 20].

In our work, we shall deal with a simple version of the model of Eq. (1), extended with a shape dependence, and discuss briefly the effect of shell corrections. The shape dependence of the binding energy comes through the functions $f_{S,C}(\alpha)$, where α are an arbitrary set of deformation parameters. We use a common expansion into spherical harmonics, restricting ourselves to the leading quadrupole term and axial shapes: $R(\theta) = R_0 \left[1 + \alpha_0 + \alpha_2 P_2(\cos\theta)\right]$. The monopole factor α_0 is restricted by the requirement of volume conservation, and dipole oscillations are identically zero in the center-of-mass coordinate system. The sign of the quadrupole parameter $\alpha_2 = \beta/\sqrt{4\pi/5}$ is crucial to distinguish an elongated (prolate) shape ($\alpha_2 > 0$) from a squeezed (oblate) shape ($\alpha_2 < 0$).

The corrections to the surface and Coulomb terms up to the 4th order in the quadrupole deformation parameter α_2 (with the volume conservation already included) are [21]

$$\begin{aligned} f_S(\alpha_2) &= 1 + \frac{2}{5}\alpha_2^2 - \frac{4}{105}\alpha_2^3 - \frac{38}{175}\alpha_2^4\ , \\ f_C(\alpha_2) &= 1 - \frac{1}{5}\alpha_2^2 - \frac{4}{105}\alpha_2^3 + \frac{157}{1225}\alpha_2^4\ . \end{aligned} \tag{2}$$

The leading quadratic term shows no difference between prolate and oblate shapes with the same deformation sizes (it demonstrates instead a competition between the surface energy, which clearly prefers a spherical shape, and the Coulomb repulsion decreasing the total energy by stretching the charged nuclear medium). Quartic terms, that are also symmetric with respect to the prolate-oblate deformation, lightly reduce the effect of quadratic terms due to their opposite signs.

It is the cubic correction that plays the crucial role in the analysis of the prolate shape predominance, since it distinguishes between prolate and oblate deformations. As follows from Eqs. (1) and (2), the prolate-oblate binding energy difference $\Delta B = B(A, Z, |\alpha_2|) - B(A, Z, -|\alpha_2|)$ takes the value

$$\Delta B = \frac{8}{105}\left[a_S A^{\frac{2}{3}} + a_C \frac{Z(Z-1)}{A^{\frac{1}{3}}}\right] |\alpha_2|^3\ . \tag{3}$$

Journal of Physics: Conference Series **322** (2011) 012018 doi:10.1088/1742-6596/322/1/012018

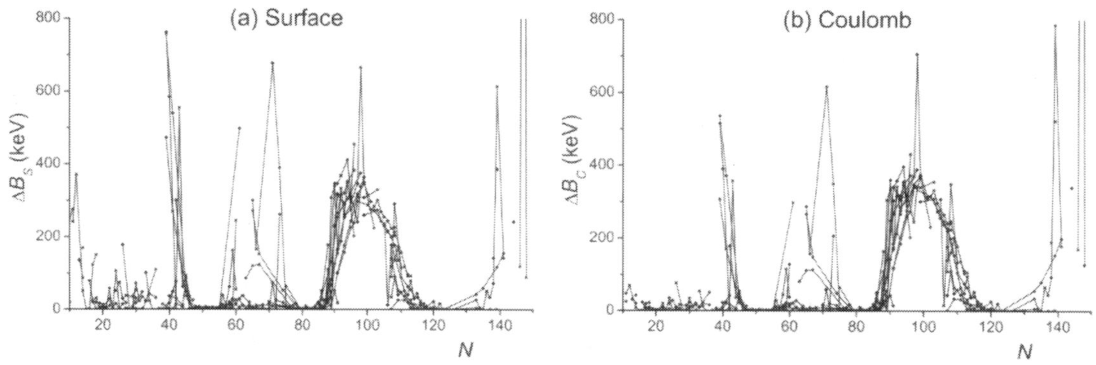

Figure 2. (a) Surface and (b) Coulomb contributions to the total prolate-oblate binding energy difference ΔB calculated from the electric quadrupole moments Q. Both contributions have very similar values, especially in the rare-earth region. Surface energy is higher for lighter nuclei, Coulomb energy becomes important with increasing charge of nuclei.

Thus, the effects of surface and Coulomb forces sum and diminish the total energy of the prolate shape, which is therefore always favored energetically over the oblate shape.

We now qualitatively discuss the effect of the shell corrections that induce shape stabilization. Numerical values of the LDM parameters a_S and a_C [see Eq. (7) below] imply that the global minimum of nuclei around the valley of stability always occurs at $\alpha_2 = 0$, i.e. for a spherical shape. In order to describe the ground state deformation, shell effects must be taken into account. Under the assumption of a bunched Fermi-gas spectrum of levels, Myers and Świątecki derived an explicit formula for the shell corrections [19] with the following properties: (i) the energy of the shell corrections is deeply negative at shell closures, exceeding a value of 10 MeV (ii) it is slightly positive in mid-shell regions, reaching values of a few MeV, and (iii) it decreases rapidly with an attenuating factor $\exp\left(-\alpha_2^2/c^2\right)$ when the shape is deformed (c is an adjustable parameter). Thus, at closed shells the spherical shape is perfectly stable. In the intermediate region, however, the fast exponential decay of positive shell correction energy competes with the rise in the collective part of the binding energy due to the quadratic terms in Eq. (2). The spherical shape is not stable anymore. Two local minima appear instead, one on the prolate side and the other on the oblate. Since the attenuating fraction is symmetric with respect to deformation, it does not contribute to the binding energy difference of (3), and therefore the prolate minimum always lies deeper in energy. Note that the appearance of few oblate nuclei can be explained by more sophisticated Strutinsky's shell corrections, as presented, for example, in [14]. However, the robustness of the Strutinsky's method can be called into question [12].

3. Results

The question can now be raised: is the difference in binding energy between prolate and oblate deformation given by the liquid drop model large enough to explain the observed predominance of deformed nuclei with prolate shapes? In order to answer this, we calculate the values of ΔB employing two distinct sets of experimental data from which the quadrupole deformation parameter β of axially symmetric nuclei can be extracted: (i) the electric quadrupole moment Q (reflecting the deviation of the charge distribution of the protons from the spherical shape), and (ii) the transition probability $B\left(E2 : 0^+ \to 2^+\right)$ of the first excited 2^+ state in even-even nuclei [8]:

$$\beta_Q = 0.95\frac{5}{4}\frac{Q_0}{ZR^2}\ , \tag{4}$$

and

$$|\beta_{\mathrm{B(E2)}}| = \frac{4\pi}{3ZR^2}\sqrt{B(E2)}\ . \tag{5}$$

Here $R = R_0A^{1/3}$ is the nuclear radius ($R_0 = 1.20$ fm) and Q_0 the body-fixed (intrinsic) electric quadrupole moment which is related to the space-fixed (spectroscopic) quadrupole moment Q of a state with spin I and projection onto the symmetry axis K:

$$Q = \frac{3K^2 - I(I+1)}{(I+1)(2I+3)}Q_0\ . \tag{6}$$

Note that Q can supply information of both magnitude and sign of deformation, whereas $B(E2)$ values give only magnitude [see Eqs. (4), (5)].

Numerical values of the LDM coefficients $a_{S,C}$ for the binding energies are usually determined in studies of nuclear masses and exhibit very little variation in the empirical values among different fits [12, 19, 20]. Since our model highly resembles the model used in Ref. [19], we use the values listed there:

$$a_S^{(0)} = 18.56\ \mathrm{MeV}, \qquad a_C = 0.72\ \mathrm{MeV}, \qquad \kappa = 1.79, \tag{7}$$

where $a_S = a_S^{(0)}\left(1 - \kappa I^2\right)$ is the surface parameter corrected to the neutron excess through an additional parameter κ.

Figure 1 shows the values of deformation parameter β that were determined from a compilation of experimental quadrupole moments [3], picking up states with $I = 2$ and $K = 0$ from the yrast band for even nuclei and $I = K \neq 1/2$ states for odd nuclei. In cases of more than one listed value of Q for a given nuclide, we selected the most recent value or that with the smallest relative error; for the sake of clarity the error is not shown. In the region of very light nuclei the value of the deformation parameter seems to be scattered with no apparent order, but from the shell closure $N = 50$ on, the figure exhibits a clear pattern: the deformation β is well pronounced inside the shells, while it tends to be much smaller for closed shells. This effect is even more obvious from panel (b), where the rare-earth nuclei are displayed in detail. This region exhibits a high correlation in the size of the deformation between isotones, and inside the shell the deformation is nearly constant with a well-known average of ≈ 0.3. A few oblate nuclei appear near shell closures, but a value $\beta < -0.2$ is rather exceptional.

The distribution of the total prolate-oblate binding energy difference between the spherical and the Coulomb part calculated with β_Q of the electric quadrupole moments is depicted in Figure 2. The ratio of both contributions, $(a_S/a_C)A/[Z(Z-1)]$, is larger than one for very light nuclei and it decreases with rising proton number. In the region of rare-earth nuclei it is ≈ 1, so that both contributions have almost the same magnitude of the order of 300-400 keV each except for nuclei near shell closures.

Figure 3 shows the combined effect of the surface and Coulomb deformation energy. It compares values of ΔB obtained (a) from known $B(E2)$ values of even-even nuclei in the whole nuclear chart [17], and (b) from electric quadrupole moments Q in the region of rare-earth nuclei. The binding energy difference between the prolate and oblate shapes reaches values up to 800 keV for a few nuclei at mid-shell, which is a significant amount, especially in comparison with typical excitation energies of the first excited state in deformed nuclei (in the rare-earth region this is typically of the order of 100-200 keV, for even-even nuclei lower than 100 keV).

Finally, Figure 4 shows the distribution of prolate-oblate energy difference values for all nuclei determined from the compilation of the Q moments. For a major part of nuclei ΔB is lower

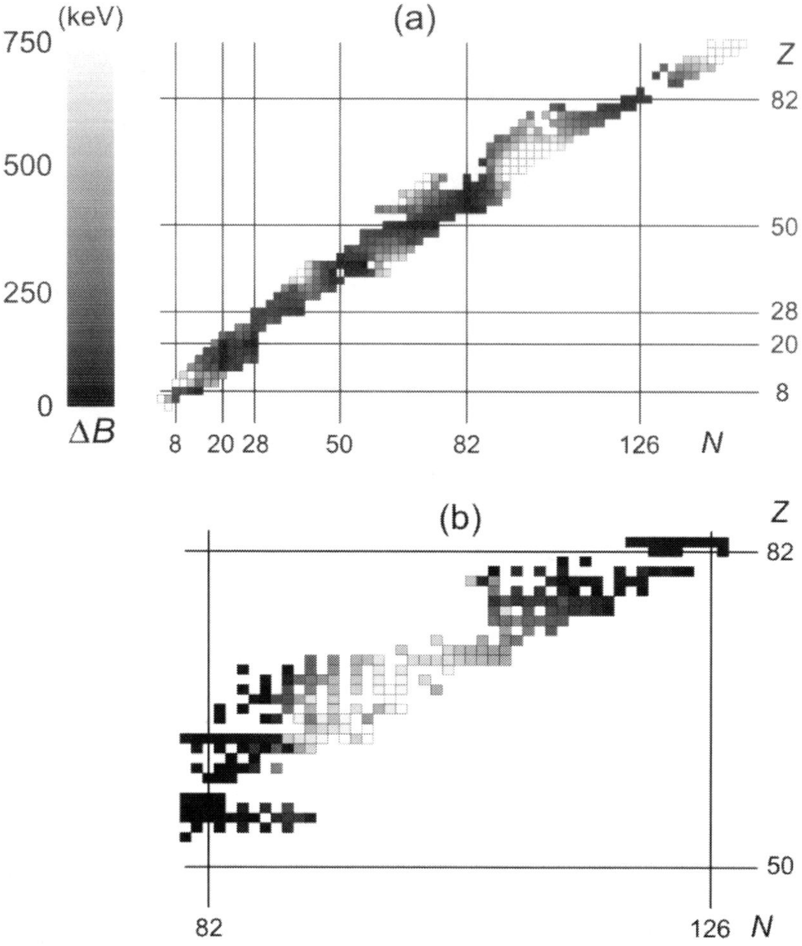

Figure 3. Total prolate-oblate binding energy difference in keV (coded in shades of gray), and calculated from the experimental values of (a) B(E2) transition probabilities (for even-even nuclei only, showing the whole nuclear chart), and (b) electric quadrupole moments Q (showing the rare-earth region only).

than 100 keV (see the first bin), which correspond to slightly deformed shapes. On the other hand, about a third of all nuclei show a difference ΔB that exceeds 100 keV, and there is even a significant amount of nuclei that have this energy difference in the range 500-1000 keV. This demonstrates that nuclei with significantly high value of δB are not an exception. Their presence in different areas of the nuclear chart give strong evidence in favor of the simple collective explanation of predominant prolate shapes.

4. Concluding remarks

In this paper we have shown that surface and Coulomb collective effects are strong enough to induce the propensity of nuclei to acquire prolate shapes. We employed the deformed liquid drop model in which the effects of quadrupole deformation are taken into account up to quartic order. It was shown that the cubic corrections to the surface and Coulomb terms contribute approximately equally to the prolate-oblate binding energy difference with the exception of very light nuclei. An analysis of the data based on compilations of experimental values of

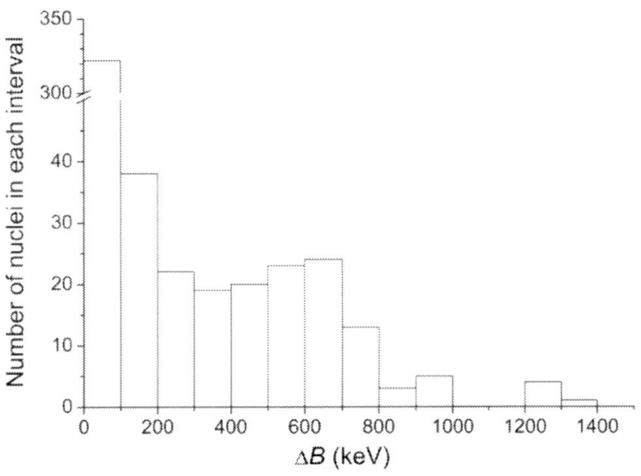

Figure 4. A histogram showing the distribution of the total prolate-oblate binding energy difference ΔB calculated from the electric quadrupole moments Q. A significant number of nuclei with ΔB of the order of several hundreds keV's is observed.

quadrupole moments and $B(E2)$ transitions yield energy differences of the order of 500-800 keV for a significant number of nuclei. These energies are comparable with typical excitation energies in deformed nuclei, and also with respect to shell corrections to the deformed LDM. The results suggest that the observed predominance of prolate nuclear shapes can be essentially explained as a collective effect with a transparent physical picture. A more detailed analysis will be presented in [6].

Acknowledgments

This work was supported in part by grants from PAPIIT-DGAPA (IN 113711), the Czech Science Foundation (202/09/0084) and the Czech Ministry of Education (MSM 0021620859).

References

[1] Hamamoto I and Mottelson B R 2009 *Phys. Rev.* C **79** 034317
[2] Horoi M and Zelevinsky V 2010 *Phys. Rev.* C **81** 034306
[3] Stone N J 2005 *At. Data Nucl. Data Tables* **90** 75
[4] Federman P and Pittel S 1977 *Phys. Lett.* B **69** 385
Federman P and Pittel S 1978 *Phys. Lett.* B **77** 29
[5] Casten R F 2000 *Nuclear Structure from a Simple Perspective* (New York: Oxford University Press)
[6] Frank A, Bijker R and Stránský P in preparation
[7] Lemmer R H and Weisskopf V F 1961 *Nucl. Phys.* **25** 624
[8] Bohr A and Mottelson B R 1998 *Nuclear Structure* (Singapore: World Scientific)
[9] Frisk H 1990 *Nucl. Phys.* A **511** 309
[10] Zickendraht W 1985 *Phys. Rev. Lett.* **54** 1906
Zickendraht W 1990 *Ann. Phys.* (Berlin) **47** 490
[11] Castel B and Goeke K 1976 *Phys. Rev.* C **13** 1765
[12] Möller P, Nix J R, Myers W D and Świątecki W J 1995 *At. Data Nucl. Data Tables* **59** 185
[13] Kumar K 1970 *Phys. Rev.* C **1** 369
[14] Tajima N and Suzuki N 2001 *Phys. Rev.* C **64** 037301
Tajima N, Shimizu Y R and Suzuki N 2002 *Prog. Theor. Phys. Suppl.* **146** 628
[15] Goriely S, Tondeur F and Pearson J M 2001 *At. Data Nucl. Data Tables* **77** 311
[16] Moshinsky M private communication

[17] Raman S, Nestor C W and Tikkanen P 2001 *At. Data Nucl. Data Tables* **78** 1
[18] Von Weizsäcker C F 1935 *Z. Phys.* **96** 431
Bethe H A and Bacher R F 1936 *Rev. Mod. Phys.* **8** 82
[19] Myers W D and Świątecki W J 1966 *Nucl. Phys.* **81** 1
[20] Pomorski K and Dudek J 2003 *Phys. Rev.* C **67** 044316
[21] Świątecki W J 1956 *Phys. Rev.* **104** 993

XXXIV Symposium on Nuclear Physics IOP Publishing
Journal of Physics: Conference Series **322** (2011) 012019 doi:10.1088/1742-6596/322/1/012019

Uncertainties in nuclear transition matrix elements for neutrinoless $\beta\beta$ decay

P K Rath

Department of Physics, University of Lucknow, Lucknow-226007, India

E-mail: pkrath_lu@yahoo.co.in

Abstract. To ascertain the uncertainties associated with the nuclear transition matrix elements $M^{(0\nu)}$ for the neutrinoless $\beta\beta$ decay of ^{94,96}Zr, ^{98,100}Mo, ^{104}Ru, ^{110}Pd, ^{128,130}Te and ^{150}Nd isotopes in the case of $0^+ \rightarrow 0^+$ transition, a statistical analysis has been performed by calculating eight (twelve) different nuclear transition matrix elements $M^{(0\nu)}$ in the projected Hartree-Fock-Bogoliobov (PHFB) model using four different parameterizations of a Hamiltonian with pairing plus multipolar effective two-body interaction and two (three) different parameterizations of Jastrow-type short range correlations. The averages in conjunction with their standard deviations provide an estimate of the uncertainties associated the nuclear transition matrix elements $M^{(0\nu)}$ calculated within the PHFB model.

1. Introduction

The observation of lepton number violating neutrinoless double beta $(\beta\beta)_{0\nu}$ decay is a convenient tool for establishing the Majorana nature of neutrinos. Once observed, it can provide information on many a gauge theoretical parameters, namely effective mass of light as well as heavy neutrinos, the effective coupling constants of left-right and right-right handed currents in the left-right symmetric models, the intergeneration Yukawa coupling constants in the R_p-violating minimal super-symmetric (SUSY) standard model, leptoquark-Higgs coupling constant, mixing parameters of heavy sterile neutrinos with light Majorana neutrinos, the compositeness scale, the brain-shift parameter of extradimensional models and Majoron-neutrino coupling constants in addition to the violation of Lorentz invariance and weak equivalence principle. According to the "black box theorem" [1], the observation of $(\beta\beta)_{0\nu}$ decay implies the non-zero mass of Majorona neutrinos at the weak scale in any gauge theoretical model with spontaneous symmetry breaking independent of underlying mechanisms. The extension of this "black box theorem" to SUSY models [2], implies the existence of massive sneutrinos provided the $(\beta\beta)_{0\nu}$ decay is observed.

The reliability of extracted gauge theoretical parameters depends on the accuracy of calculated model dependent nuclear transition matrix elements (NTMEs). The nuclear many-body theory is quite complex as it deals with the nonperturbative region of quantum chromo dynamics (QCD). Hopefully, the present attempts to develop the effective field theory (EFT) as an alternative method for low energy QCD will emerge as a powerful technique for solving nuclear many-body theory in the near future. The basic problem in the conventional nuclear structure calculations is to solve the nuclear many body problem perturbatively to all orders and the nuclear shell model is a prototypical nuclear many-body theory. However, most of the $\beta\beta$ emitters are medium and heavy mass nuclei for which there is a necessity of large scale

Journal of Physics: Conference Series **322** (2011) 012019 doi:10.1088/1742-6596/322/1/012019

configuration mixing and the presently available computational facilities are limited for this purpose. Therefore, one is forced to look for alternative models such as quasiparticle random-phase approximation (QRPA) and their extensions, projected-Hartree-Fock-Bogoliobov model (PHFB), deformed Hartree-Fock (DHF) and generator coordinate method plus particle number and angular momentum projection (GCM+PNAMP), single state dominance hypothesis (SSDH) as well as group theoretical methods.

In general, there are four main sources of uncertainties, namely the inclusion of pseudoscalar and weak magnetism terms in the Fermi, Gamow-Teller (GT) and tensorial NTMEs [3, 4], the effects due to finite size of nucleons (FNS) as well as short range correlations (SRC) [5, 6, 7, 8, 9, 10, 11, 12], deformation and the use of two effective values of the axial-vector coupling constant g_A. In addition, different nuclear models produce different NTMEs even for a given transition depending on the approximations involved and for a given model, NTMEs also depend on the model space and effective two-body interaction selected. Based on these observations, Vogel proposed that the spread between the calculated NTMEs can provide a measure of the theoretical uncertainty [13]. In the case of the well studied ^{76}Ge isotope, it was observed that the calculated decay rates $T^{0\nu}_{1/2}$ differ by a factor of 6–7 and hence, the uncertainty in the effective neutrino mass $\langle m_\nu \rangle$ is about 2–3. Experimental limit $T^{0\nu}_{1/2} > 1.6 \times 10^{25}$ yr [14] imply upper limits on $\langle m_\nu \rangle$ between 0.4 eV and 1.0 eV, depending on the NTMEs [15, 16, 17].

By performing a statistical analysis, the spread between the calculated NTMEs can be translated into averages and standard deviations [18, 19]. According to Bilenky and Grifols [20], the comparison of calculated ratios of the corresponding NTMEs-squared and the ratios of half-lives could test the validity of nuclear structure calculations in a model independent way once the $(\beta\beta)_{0\nu}$ decay would be observed in several nuclei. Rodin *et al.* [21] have estimated the theoretical uncertainty employing QRPA and renormalized QRPA (RQRPA), with three sets of basis states and three realistic two-body effective interactions. Different strategies to remove the sensitivity of QRPA calculations on the model parameters have been also proposed [22, 23]. Further studies on uncertainties in NTMEs due to SRC using the unitary correlation operator method (UCOM) [10] and self-consistent coupled cluster method (CCM) [11] have been carried out by Tuebingen group.

Specifically, the PHFB model is unique in allowing the description of the $\beta\beta$ decay in medium and heavy mass nuclei by projecting a set of states with good angular momentum, while treating the pairing and deformation degrees of freedom simultaneously and on equal footing. The PHFB model, in conjunction with pairing plus quadrupole-quadrupole (PQQ) interaction [24] has been successful in the study of the $0^+ \rightarrow 0^+$ transition of $(\beta^-\beta^-)_{2\nu}$ decay, and it was possible to describe the lowest excited states of the parent and daughter nuclei along with their electromagnetic transition strengths, as well as to reproduce the measured $\beta\beta$ decay rates [25, 26]. However, the structure of the intermediate odd Z-odd N nuclei and hence, the single β decay rates and the distribution of GT strength can not be studied in the present version of the PHFB model. In spite of this limitation, it is quite convenient for examining the explicit role of deformation on the NTMEs [27, 28]. The effects of pairing and quadrupolar correlations on the NTMEs of $(\beta^-\beta^-)_{0\nu}$ decay has been also studied in the interacting shell model (ISM) [9, 29].

Presently, two different parameterizations of the QQ interaction have been employed, with and without the HH correlations. Further, the NTMEs $M^{(0\nu)}$ are calculated with three different parametrizations of Jastrow SRC employing the four sets of wave functions. The twelve NTMEs provide a reasonable sample for estimating the associated uncertainties. In Sec. 2, the reliability of HFB wave functions has been discussed briefly. In Sec. 3, the effects due to the four different parameterizations of the pairing plus multipole Hamiltonian and three different parametrizations of SRC on the calculated NTMEs are analyzed, and in Sec. 4, their average values as well as standard deviations are estimated. Subsequently, the latter are employed to obtain upper limits on the effective mass of light Majorana neutrinos. Conclusions are given in Sec. 5.

Journal of Physics: Conference Series **322** (2011) 012019 doi:10.1088/1742-6596/322/1/012019

2. Reliability of HFB wave functions

The HFB wave functions were generated using an effective Hamiltonian [28] given by

$$H = H_{sp} + V(P) + V(QQ) + V(HH), \tag{1}$$

where H_{sp}, $V(P)$, $V(QQ)$ and $V(HH)$ denote the single particle Hamiltonian, the pairing, quadrupole-quadrupole (QQ) and hexadecapole-hexadecapole (HH) parts of the effective two-body interaction, respectively. The model space and single particle energies (SPE's) have been already given in Refs. [25, 26, 27]. In the quadrupole-quadrupole part of the effective two-body interaction, the $V(QQ)$ has three terms for proton-proton, neutron-neutron and the proton-neutron interaction with coefficients χ_{2pp}, χ_{2nn} and χ_{2pn}, respectively. The strengths of the like particle components of the QQ interaction were taken as $\chi_{2pp} = \chi_{2nn} = 0.0105$ MeV b^{-4}. By fitting the experimental excitation energy of the 2^+ state, E_{2^+}, the strength of proton-neutron component of the QQ interaction χ_{2pn} was fixed [25, 26, 27]. Alternatively, one can also employ an isoscalar parametrization of the quadrupole-quadrupole interaction, by taking $\chi_{2pp} = \chi_{2nn} = \chi_{2pn}/2$, and the three parameters were varied together to fit E_{2^+}. These two parameterizations of the quadrupole-quadrupole interaction are refereed as $PQQ1$ and $PQQ2$ with PQQ [24] type of effective two-body interaction.

The experimental excitation energies of 2^+ state E_{2^+} [30] can be reproduced within about 2% accuracy in both methods. By employing the $PQQ1$ parametrization, the maximum change in E_{4^+} and E_{6^+} energies with respect to $PQQ1$ interaction [25, 26] is about 5% and 18%, respectively. By employing either parametrizations, the reduced $B(E2{:}0^+ \to 2^+)$ transition probabilities, deformation parameters β_2, static quadrupole moments $Q(2^+)$ and gyromagnetic factors $g(2^+)$ are in an overall agreement with the experimental data [31, 32]. With respect to $PQQ1$ parametrization, the maximum change in the calculated NTMEs $M_{2\nu}$ for the $0^+ \to 0^+$ transition in the case of $PQQ2$ parametrization, is about 21% but for ^{94}Zr isotope. By including the hexadecapolar term HH, we end up with four different parameterizations of the effective two-body interaction, namely $PQQ1$, $PQQHH1$, $PQQ2$ and $PQQHH2$.

3. Short range correlations and NTMEs

Considering the finite size of nucleons, the inverse half-life of the $(\beta^-\beta^-)_{0\nu}$ decay due to the exchange of light Majorana neutrinos for the $0^+ \to 0^+$ transition is given by [5, 15, 33]

$$\left[T_{1/2}^{0\nu}(0^+ \to 0^+)\right]^{-1} = \left(\frac{\langle m_\nu\rangle}{m_e}\right)^2 G_{01}|M^{(0\nu)}|^2, \tag{2}$$

and the model dependent NTME $M^{(0\nu)}$ is given by

$$M^{(0\nu)} = \sum_{n,m}\left\langle 0_F^+ \left\| \left[\sigma_n\cdot\sigma_m - \left(\frac{g_V}{g_A}\right)^2\right] H(r_{12})\tau_n^+\tau_m^+ \right\| 0_I^+ \right\rangle, \tag{3}$$

where

$$H(r_{12}) = \frac{4\pi R}{(2\pi)^3}\int d^3q \frac{\exp(i\mathbf{q}\cdot\mathbf{r_{12}})}{q\left(q+\overline{A}\right)}\left(\frac{\Lambda^2}{\Lambda^2+q^2}\right)^4, \tag{4}$$

with $\overline{A} = \langle E_N\rangle - \frac{1}{2}(E_I + E_F)$. The cutoff momentum $\Lambda =$ 850 MeV [27].

Recently, Šimkovic *et al.* [11] have shown that it is possible to parametrize the SRC effects of Argonne V18 and CD-Bonn two nucleon potentials by the Jastrow type of correlations within a few percent accuracy. Explicitly, the effects due to the SRC can be incorporated in the calculation of $M^{(0\nu)}$ through the prescription $O_k \to fO_kf$, where

$$f(r) = 1 - ce^{-ar^2}(1 - br^2), \tag{5}$$

with $a = 1.1$, 1.59 and 1.52 fm^{-2}, $b = 0.68$, 1.45 and 1.88 fm^{-2} and $c = 1.0$, 0.92 and 0.46 for Miller-Spencer [34], Argonne V18 and CD-Bonn NN potentials, respectively. Presently, the NTMEs $M^{(0\nu)}$ are calculated in the PHFB model by employing these three sets of parameters for the SRC, denoted as SRC1, SRC2 and SRC3, respectively. The three functions $f(r)$ have similar forms, but differ in their values at the origin, and at the position of their maximum, which lie at 1.54, 1.15 and 1.09 fm for SRC1, SRC2 and SRC3, respectively. As discussed below, they have clear influence on the calculated NTMEs and radial evolution of the $(\beta^-\beta^-)_{0\nu}$ decay matrix elements.

We present the twelve NTMEs $M^{(0\nu)}$ in Table 1, evaluated using the HFB wave functions in conjunction with $PQQ1$, $PQQHH1$, $PQQ2$, $PQQHH2$ interactions and three different parametrizations of the Jastrow type of SRC for the nuclei ^{94,96}Zr, ^{98,100}Mo, ^{104}Ru, ^{110}Pd, ^{128,130}Te and ^{150}Nd. Following Haxton's prescription [15], the average energy denominator $\overline{A}$ has been taken as $\overline{A} = 1.12A^{1/2}$ MeV . In Ref. [12], the NTMEs were calculated in the approximations of point nucleons (P), FNS, point nucleons with SRC (P+SRC), and finite size plus SRC (FNS+SRC). The NTMEs $M^{(0\nu)}$ were also calculated for $\overline{A}/2$ in the energy denominator in the case of point nucleons to obtain information on the validity of closure approximation. The following observations are worth mentioning [12].

(i) It was observed that the relative change in NTMEs $M^{(0\nu)}$ is of the order of 10 %, when the energy denominator was taken as $\overline{A}/2$ instead of $\overline{A}$. It confirms that there is a weak dependence of NTMEs on average excitation energy $\overline{A}$ for the $(\beta^-\beta^-)_{0\nu}$ decay and the closure approximation is quite valid.

(ii) Due to the different parameterizations of the Hamiltonians, the variation in $M^{(0\nu)}$ lies between 20–25%. In general, it was noticed that the NTMEs but for ^{128}Te isotope evaluated for both parameterizations of the quadrupolar interaction are quite close. The inclusion of the hexadecapolar term tends to reduce them by amounts which strongly depend on the specific nuclei.

(iii) In the approximation of point nucleons, the inclusion of SRC induces an extra quenching in the NTMEs $M^{(0\nu)}$, which is of the order of 18–23% for SRC1 and negligible for SRC3.

(iv) The dipole form factor always reduces the NTMEs by 12–15% in comparison to the point nucleon case.

(v) Adding SRC can further reduce the transition matrix elements, for SRC1, or slightly enhance them, partially compensating the effect of the dipole form factor. It was interesting to note that the effect of FNS+SRC2 is almost negligible, i.e., nearly the same as FNS.

The radial evolution of $M^{(0\nu)}$ has been studied by Šimkovic *et al.* [10] by defining

$$M^{(0\nu)} = \int C^{(0\nu)}(r)\,dr. \tag{6}$$

In the QRPA by Šimkovic *et al.* [10] and ISM by Menéndez *et al.* [35], it has been observed that the contributions of decaying pairs coupled to $J = 0$ and $J > 0$ almost cancel beyond $r \approx 3$ fm and the magnitude of $C^{(0\nu)}$ for all nuclei undergoing $(\beta^-\beta^-)_{0\nu}$ decay are the maximum about the internucleon distance $r \approx 1$ fm.

The study of radial dependence of $C^{(0\nu)}$ in the PHFB model due to $PQQ1$ parametrization of the effective two body interaction for six nuclei, namely ^{96}Zr, ^{100}Mo, ^{110}Pd, ^{128,130}Te and ^{150}Nd [12], provide following observations.

(i) In case of point nucleons, it was noticed that the $C^{(0\nu)}$ are peaked at $r = 1.0$ fm and with the addition of SRC1, the peak shifts to 1.25 fm. However, the magnitude of $C^{(0\nu)}$ are increased for SRC2 and SRC3 with unchanging peak position.

Journal of Physics: Conference Series **322** (2011) 012019 doi:10.1088/1742-6596/322/1/012019

Table 1. Calculated NTMEs $M^{(0\nu)}$ in the PHFB model with four different parameterizations of the effective two-body interaction, three different parameterizations of SRC along with average NTMEs $\overline{M}^{(0\nu)}$ and uncertainties $\Delta\overline{M}^{(0\nu)}$ for the $(\beta^-\beta^-)_{0\nu}$ decay of ^{94,96}Zr, ^{98,100}Mo, ^{104}Ru, ^{110}Pd, ^{128,130}Te and ^{150}Nd isotopes.

Nuclei		$M^{(0\nu)}$			g_A	$\overline{M}^{(0\nu)}$	
		SRC1	SRC2	SRC3		Case I	Case II
^{94}Zr	$PQQ1$	4.0690	4.6639	4.8383	1.254	4.2464±0.3883	4.4542±0.2536
	$PQQHH1$	3.7315	4.2820	4.4441	1.0	4.6382±0.4246	4.8668±0.2759
	$PQQ2$	3.9802	4.4818	4.6259			
	$PQQHH2$	3.5424	4.0708	4.2266			
^{96}Zr	$PQQ1$	2.9068	3.3590	3.4923	1.254	3.1461±0.2778	3.3181±0.1243
	$PQQHH1$	2.8507	3.3192	3.4578	1.0	3.4481±0.3085	3.6376±0.1424
	$PQQ2$	2.7758	3.2103	3.3385			
	$PQQHH2$	2.6745	3.1182	3.2497			
^{98}Mo	$PQQ1$	6.7322	7.6297	7.8884	1.254	7.1294±0.6013	7.4656±0.3635
	$PQQHH1$	6.1984	7.0618	7.3114	1.0	7.8398±0.6826	8.2099±0.4358
	$PQQ2$	6.7630	7.6695	7.9307			
	$PQQHH2$	6.1344	6.9925	7.2406			
^{100}Mo	$PQQ1$	6.5036	7.4282	7.6920	1.254	6.8749±0.6855	7.2163±0.4977
	$PQQHH1$	6.1597	7.0654	7.3248	1.0	7.5660±0.7744	7.9419±0.5769
	$PQQ2$	6.5534	7.4838	7.7493			
	$PQQHH2$	5.5520	6.3756	6.6113			
^{104}Ru	$PQQ1$	4.6942	5.3989	5.5975	1.254	4.8464±0.4840	5.1004±0.3280
	$PQQHH1$	4.2809	4.9548	5.1454	1.0	5.3599±0.5533	5.6396±0.3926
	$PQQ2$	4.4137	5.0777	5.2647			
	$PQQHH2$	3.9648	4.5931	4.7708			
^{110}Pd	$PQQ1$	7.6982	8.7783	9.0850	1.254	7.8413±0.8124	8.2273±0.6167
	$PQQHH1$	6.3963	7.3535	7.6262	1.0	8.6120±0.9184	9.0370±0.7128
	$PQQ2$	7.3842	8.4187	8.7128			
	$PQQHH2$	6.7982	7.7816	8.0621			
^{128}Te	$PQQ1$	3.2499	3.7258	3.8639	1.254	4.0094±0.4194	4.2175±0.3074
	$PQQHH1$	3.4994	4.0740	4.2401	1.0	4.4281±0.4601	4.6571±0.3355
	$PQQ2$	3.8893	4.4374	4.5956			
	$PQQHH2$	3.7336	4.3172	4.4860			
^{130}Te	$PQQ1$	4.4319	5.0103	5.1753	1.254	4.4458±0.5231	4.6633±0.4269
	$PQQHH1$	3.6277	4.1964	4.3595	1.0	4.9065±0.5837	5.1459±0.4802
	$PQQ2$	4.3610	4.9320	5.0951			
	$PQQHH2$	3.6218	4.1879	4.3503			
^{150}Nd	$PQQ1$	3.2316	3.6375	3.7514	1.254	3.1048±0.4649	3.2431±0.4434
	$PQQHH1$	2.4208	2.7447	2.8359	1.0	3.4334±0.5181	3.5856±0.4952
	$PQQ2$	3.1574	3.5546	3.6661			
	$PQQHH2$	2.5031	2.8311	2.9234			

Journal of Physics: Conference Series **322** (2011) 012019 doi:10.1088/1742-6596/322/1/012019

(ii) In the case of FNS, the $C^{(0\nu)}$ are peaked at $r = 1.25$ fm, which remains unchanged with the inclusion of SRC1, SRC2 and SRC3. However, the magnitudes of $C^{(0\nu)}$ change in the latter three cases.

The above observations also remain valid with the other three parametrizations of the effective two-body interaction.

4. Uncertainties in NTMEs

We have performed a statistical analysis for estimating the uncertainties associated with the NTMEs $M^{(0\nu)}$ for $(\beta^-\beta^-)_{0\nu}$ decay calculated in the PHFB model by evaluating the mean and the standard deviation, defined as

$$\overline{M}^{(0\nu)} = \frac{\sum_{i=1}^{N} M_i^{(0\nu)}}{N} \tag{7}$$

and

$$\Delta\overline{M}^{(0\nu)} = \frac{1}{\sqrt{N-1}}\left[\sum_{i=1}^{N}\left(\overline{M}^{(0\nu)} - M_i^{(0\nu)}\right)^2\right]^{1/2}. \tag{8}$$

Based on the observation by Šimkovic *et al.* [11] that the Miller-Spenser parametrization of the Jastrow correlation is a major source of uncertainty and it is better to consider SRC2 or SRC3 due to the Argonne V18 and CD-Bonn NN potentials, respectively, we have calculated the averages and variances for two cases. In case I, the statistical analysis has been performed for twelve NTMEs listed in Table 1, for the bare and quenched values of axial vector coupling constant $g_A = 1.254$ and $g_A = 1.0$, respectively. Similarly, the averages and standard deviations are calculated for eight NTMEs $M^{(0\nu)}$ due to SRC2 and SRC3 in case II. In Table 1, the averages $\overline{M}^{(0\nu)}$ and uncertainties $\Delta\overline{M}^{(0\nu)}$ have been displayed. It is noticed that the exclusion of Miller-Spencer parametrization reduces the uncertainty by about 55% in ^{96}Zr to 4% in ^{150}Nd.

In Fig. 1, we have plotted the average NTMEs $\overline{M}^{(0\nu)}$ with the uncertainties $\Delta\overline{M}^{(0\nu)}$ along with the recently available results obtained by employing the ISM [9], QRPA, RQRPA [11], IBM [36] and GCM+PNAMP [37]. It is observed that in all the models, the NTMEs $M^{(0\nu)}$ have a weak or no dependence on the mass number A. Further, upper limits on the effective mass of light neutrinos $\langle m_\nu\rangle$ have been extracted from the largest observed limits on half-lives $T_{1/2}^{0\nu}$ of $(\beta^-\beta^-)_{0\nu}$ decay [38, 39, 40, 41, 42, 43, 44] using the phase space factors of Boehm and Vogel [45]. The extracted limits on $\langle m_\nu\rangle$ for ^{100}Mo and ^{130}Te nuclei are $0.51^{+0.06}_{-0.05} - 0.73^{+0.08}_{-0.07}$ eV and $0.31^{+0.04}_{-0.03} - 0.45^{+0.06}_{-0.05}$ eV, respectively. Assuming $\langle m_\nu\rangle = 50$ meV, the predicted half-lives of $(\beta^-\beta^-)_{0\nu}$ decay of ^{96}Zr, ^{100}Mo, ^{128,130}Te and ^{150}Nd isotopes for $g_A = 1.254(1.0)$ are $1.60^{+0.13}_{-0.11}\times10^{26}$ ($3.29^{+0.27}_{-0.24}\times10^{26}$) yr, $4.32^{+0.66}_{-0.54}\times10^{25}$ ($8.83^{+1.44}_{-1.15}\times10^{25}$) yr, $3.18^{+0.52}_{-0.42}\times10^{27}$ ($6.44^{+1.04}_{-0.84}\times10^{27}$) yr, $1.07^{+0.23}_{-0.17}\times10^{26}$ ($2.17^{+0.47}_{-0.35}\times10^{26}$) yr and $4.69^{+1.60}_{-1.06}\times10^{25}$ ($9.49^{+3.29}_{-2.16}\times10^{25}$) yr, respectively.

5. Conclusions

In our previous works [25, 26, 28], the reliability of wave functions generated with $PQQ1$ and $PQQHH1$ interactions was tested by calculating the yrast spectra, reduced $B(E2{:}0^+ \to 2^+)$ transition probabilities, static quadrupole moments $Q(2^+)$ and g-factors $g(2^+)$ of participating nuclei in $(\beta^-\beta^-)_{2\nu}$ decay as well as $M_{2\nu}$ and comparing them with the available experimental data. It was observed that the PHFB wave functions generated by fixing χ_{pn} or χ_{pp} to reproduce the E_{2^+} are reasonably reliable in reproducing an overall agreement between the calculated and observed spectroscopic properties as well as $M_{2\nu}$.

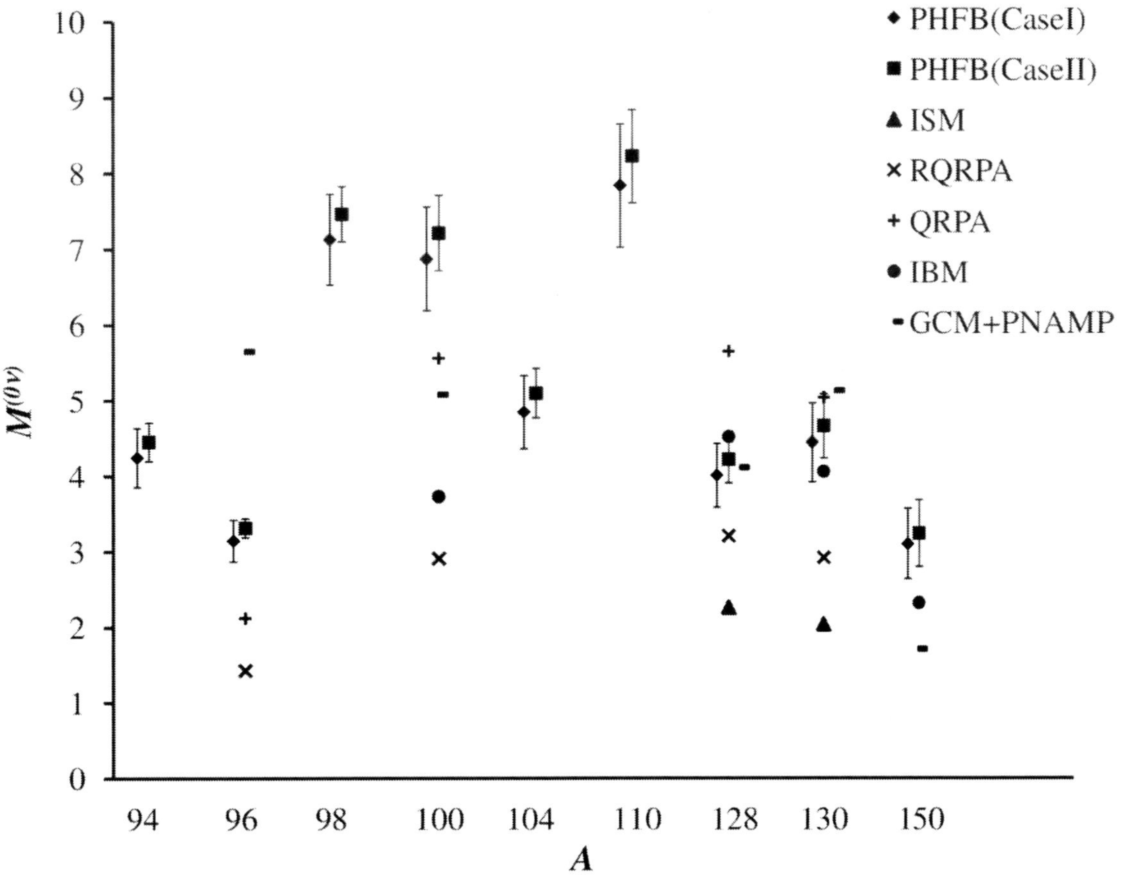

Figure 1. NTMEs calculated in ISM [9], QRPA, RQRPA [11], PHFB [12], IBM [36] and GCM+PNAMP [37] in the mass range $A = 90 - 150$.

Presently, eight (twelve) NTMEs $M^{(0\nu)}$ for $(\beta^-\beta^-)_{0\nu}$ decay of ^{94,96}Zr, ^{98,100}Mo, ^{104}Ru, ^{110}Pd, ^{128,130}Te and ^{150}Nd isotopes were calculated by employing four different parameterizations of the pairing plus multipolar type of effective two body interaction, FNS and two (three) different parameterizations of Jastrow SRC. The mean and standard deviations were evaluated for the eight (twelve) NTMEs $M^{(0\nu)}$ for both $g_A = 1.254$ and $g_A = 1.0$ to estimate the uncertainties in the NTMEs $M^{(0\nu)}$ calculated in the PHFB model. The largest theoretical uncertainty, turns out to be of the order of 15% in the case of ^{150}Nd isotope. Also, limits have been extracted on the effective mass of light Majorana neutrinos $\langle m_\nu \rangle$ from the available limits on experimental half-lives $T_{1/2}^{0\nu}$ using average NTMEs $\overline{M}^{(0\nu)}$. In the case of ^{130}Te isotope, one obtains the best limit on the effective neutrino mass $\langle m_\nu \rangle < 0.31^{+0.04}_{-0.03} - 0.45^{+0.06}_{-0.05}$ eV from the observed limit on the half-lives $T_{1/2}^{0\nu} > 3.0 \times 10^{24}$ yr of $(\beta^-\beta^-)_{0\nu}$ decay [43].

Acknowledgments
The author thanks his collaborators Drs. J. G. Hirsch, UNAM, Mexico, P. K. Raina and R. Chandra, IIT, Kharagpur and K. Chaturvedi, Bundelkhand University, India for valuable scientific discussions, organizers of XXXIV Symposium on Nuclear Physics, Cocoyoc, Mexico

for providing the opportunity to participate in an interactive and well organized symposium and CSIR, CICS, India as well as UNAM, Mexico for financial assistance to participate in the above mentioned Symposium. This work has been partially supported by DST, India vide sanction No. SR/S2/HEP-13/2006 and DST-RFBR Collaboration via grant No. RUSP-935.

References

[1] Schechter J and Valle J W F 1982 *Phys. Rev.* D **25** 2951
[2] Hirsch M, Klapdor-Kleingrothaus H V and Kovalenko S G 1997 *Phys. Lett.* B **398** 311
[3] Šimkovic F, Pantis G, Vergados J D and Faessler A 1999 *Phys. Rev.* C **60** 055502
[4] Vergados J D 2002 *Phys. Rep.* **361** 1
[5] Tomoda T 1991 *Rep. Prog. Phys.* **54** 53
[6] Wu H F, Song H Q, Kuo T T S, Cheng W K and Strottman D 1985 *Phys. Lett.* B **162** 227
[7] Hirsch J G, Castaños O and Hess P O 1995 *Nucl. Phys.* **A582** 124
[8] Kortelainen M and Suhonen J 2007 *Phys. Rev.* C **76** 024315; Kortelainen M, Civitarese O, Suhonen J and Toivanen J 2007 *Phys. Lett.* B **647** 128
[9] Caurier E, Menéndez J, Nowacki F and Poves A 2008 *Phys. Rev. Lett.* **100** 052503
[10] Šimkovic F, Faessler A, Rodin V, Vogel P and Engel J 2008 *Phys. Rev.* C **77** 045503
[11] Šimkovic F, Faessler A, Müther H, Rodin V and Stauf M 2009 *Phys. Rev.* C **79** 055501
[12] Rath P K, Chandra R, Chaturvedi K, Raina P K and Hirsch J G 2010 *Phys. Rev.* C **82** 064310
[13] Vogel P 2001 *Current Aspects of Neutrino Physics* edited by Caldwell D O (Springer Verlag, New York)
[14] Aalseth C E *et al.* 2002 *Phys. Rev.* D **65** 092007; 2004 *Phys. Rev.* D **70** 078302
[15] Haxton W C and Stephenson Jr. G J 1984 *Prog. Part. Nucl. Phys.* **12** 409
[16] Engel J, Vogel P and Zirnbauer M R 1988 *Phys. Rev.* C **37** 731
[17] Staudt A, Muto K and Klapdor-Kleingrothaus H V 1990 *Europhys. Lett.* **13** 31
[18] Bahcall J N, Murayama H and Peña-Garay C 2004 *Phys. Rev.* D **70** 033012
[19] Avignone F T, Elliott S R and Engel J 2008 *Rev. Mod. Phys.* **80** 481
[20] Bilenky S M and Grifols J A 2002 *Phys. Lett.* B **550** 154
[21] Rodin V, Faessler A, Šimkovic F and Vogel P 2003 *Phys. Rev.* C **68** 044302
[22] Suhonen J 2005 *Phys. Lett.* B **607** 87
[23] Rodin V A, Faessler A, Šimkovic F and Vogel P 2006 *Nucl. Phys.* **A766**, 107; 2007 *Nucl. Phys.* **A793**, 213
[24] Baranger M and Kumar K 1968 *Nucl. Phys.* **A110** 490
[25] Chandra R, Singh J, Rath P K, Raina P K and Hirsch J G 2005 *Eur. Phys. J.* A **23** 223
[26] Singh S, Chandra R, Rath P K, Raina P K and Hirsch J G 2007 *Eur. Phys. J.* A **33** 375
[27] Chaturvedi K, Chandra R, Rath P K, Raina P K and Hirsch J G 2008 *Phys. Rev.* C. **78** 054302
[28] Chandra R, Chaturvedi K, Rath P K, Raina P K and Hirsch J G 2009 *Europhys. Lett.* **86** 32001
[29] Caurier E, Nowacki F and Poves A 2008 *Eur. Phys. J.* A **36** 195
[30] Sakai M 1984 *At. Data Nucl. Data Tables* **31** 399
[31] Raman S, Nestor Jr. C W and Tikkanen P 2001 *At. Data Nucl. Data Tables* **78** 1
[32] Raghavan P 1989 *At. Data Nucl. Data Tables* **42** 189
[33] Doi M, Kotani T and Takasugi E 1985 *Prog. Theor. Phys. Suppl.* **83** 1
[34] Miller G A and Spencer J E 1976 *Ann. Phys. (NY)* **100** 562
[35] Menéndez J, Poves A, Caurier E and Nowacki F 2009 *Nucl. Phys.* **A818** 139
[36] Barea J and Iachello F 2009 *Phys. Rev.* C **79** 044301
[37] Rodríguez T R and Pinedo G M 2010 *Phys. Rev. Lett.* **105** 252503
[38] Arnold R *et al.* 1999 *Nucl. Phys.* **A658** 299
[39] Fremlin J H and Walters M C 1952 *Proc. Phys. Soc. Lond.* A **65** 911
[40] Arnold R *et al.* 2005 *Phys. Rev. Lett.* **95** 182302
[41] Winter R G 1952 *Phys. Rev.* **85** 687
[42] Arnaboldi C *et al.* 2003 *Phys. Lett.* B **557** 167
[43] Arnaboldi C *et al.* 2008 *Phys. Rev.* C **78** 035502
[44] Argyriades J *et al.* 2009 *Phys. Rev.* C **80** 032501(R)
[45] Boehm F and Vogel P 1992 *Physics of Massive Neutrinos*, 2nd ed. (Cambridge University Press, Cambridge, 1992) p 163

AUTHOR INDEX